COMMERCIAL AND SUBLIME

Science and Culture in the Nineteenth Century

Bernard Lightman, Editor

Commercial and Sublime

POPULAR ASTRONOMY LECTURES IN BRITAIN, 1780–1860

Hsiang-Fu Huang

UNIVERSITY OF PITTSBURGH PRESS

Published by the University of Pittsburgh Press, Pittsburgh, Pa., 15260
Copyright © 2025, University of Pittsburgh Press
All rights reserved
Manufactured in the United States of America
Printed on acid-free paper
10 9 8 7 6 5 4 3 2 1

Cataloging-in-Publication data is available from the Library of Congress

Hardcover: 978-0-8229-4873-5
Paperback: 978-0-8229-6781-1

Cover art: The Proscenium of the English Opera House in the Strand (Late Lyceum). V&A: S.176-1997. © Victoria and Albert Museum, London.
Cover design: Melissa Dias-Mandoly

Publisher: University of Pittsburgh Press, 7500 Thomas Blvd., 4th floor, Pittsburgh, PA 15260, United States, www.upittpress.org
EU Authorized Representative: Easy Access System Europe, Mustamäe tee 50, 10621 Tallinn, Estonia, gpsr.requests@easproject.com

To my parents, sister, and wife

Contents

Acknowledgments
xi

Note on Currency
xv

Introduction:
A Sublime and Commercial Business
3

Chapter 1.
Pioneers
19

Chapter 2.
Geography
37

Chapter 3.
Affiliation
81

Chapter 4.
Subjects
111

Chapter 5.
Apparatus
151

Chapter 6.
Audiences
191

Epilogue:
An Enduring Legacy
225

Appendix: Astronomical Lecturers in Britain, 1800–1870
237

Notes
241

Bibliography
273

Index
305

Acknowledgments

Writing acknowledgments is the last step in completing this book. It is also a good time to reflect on the journey I have taken. This journey began at the University College London (UCL) and culminated in the publication of this book. It marks the completion of a long-standing goal and serves as a milestone for the first decade of my academic career. Along the way, I have been deeply grateful to many people who assisted me in my research and supported the completion of this monograph.

First and foremost, tribute goes to my supervisors at UCL, Joe Cain and Simon Werrett, who made enormous efforts to guide my research. Joe and Simon were patient and supportive of me as an early career academic, even during the rocky road of my often unsuccessful job search. I enjoyed and still miss my student years and the warm discussions we shared on Gower Street. I also wish to extend special thanks to Jane Gregory. This research would not have begun without Jane's vision.

I am grateful as well to Frank James and Richard Bellon, who provided invaluable advice and feedback for my research. Frank's research style, while not fancy, embodied the dedication and seriousness of a historian studying the history of the Royal Institution. Even after I shifted my research focus away from nineteenth-century British history, Richard's amiable manners and versatile research interests have continued to inspire me. I learned a great deal from these two gentlemen.

Bernard Lightman has been the patron saint of this book. His vast knowledge, boundless kindness, and precious encouragement helped

ACKNOWLEDGMENTS

me endure the journey of refining this monograph. I am grateful to him for inviting me to submit my manuscript to the University of Pittsburgh Press series Science and Culture in the Nineteenth Century. It has been a privilege to have Bernie as the series editor for this book. His unrelenting academic enthusiasm and generous encouragement to young scholars have set an example that I aspire to emulate one day.

During my research and as I wrote this book, many people offered their professional insights, commented on drafts, and provided useful information. I especially thank Geoffrey Cantor, who kindly explained to me the religious context in nineteenth-century British science and greatly improved my manuscript. I fondly remember our intellectual conversations over coffee in Bloomsbury. I also thank Martin Bush, whose energetic and lighthearted conversation polished the chapter on orreries and lantern slides. Martin has become an important collaborator in my subsequent research, and I am lucky to benefit from his expert expertise on astronomical visual aids. I would like to thank Jean-Baptiste Gouyon, Charlotte Sleigh, Rebekah Higgitt, Don Leggett, Melanie Keene, Jonathan Topham, James Elwick, William Macauley, David A. Kirby, Silvia De Bianchi, Cristiano Turbil, Michael Kliegl, Thomas Rose, Andy Gregory, and Steve Miller. Special thanks go to Iwan Morus, who kindly supplied many helpful comments when I submitted my first research article to the journal *History of Science*. I also appreciate many scholars with whom I had fruitful conversations as they shared their ideas, including Robert Smith, Jan Golinski, Charlotte Bigg, Pedro Raposo, Kurt Vanhoutte, Joshua Nall, James Sumner, Andreas Sommer, Ke Zunke, Peter Bowler, and James Secord. Among them, Robert, Jan, Charlotte, and Pedro organized a panel with me in the Three Societies Meeting at Edmonton, Canada, in 2016, which I enjoyed very much.

The UCL Department of Science and Technology Studies was my academic home when I conducted research and started to conceive this book. It provided the important resources and support I needed throughout my research. I am grateful to the department for support my work by appointing me as Honorary Research Associate. For making the department such an enjoyable and accommodating place, I especially thank Jon Agar and Chiara Ambrosio, and Alastair Tatam, whose administrative support was always a delight. I am also thankful to many friends and colleagues in the department, especially Melanie

ACKNOWLEDGMENTS

Smallman, Simon J. Lock, Hattie Lloyd, Rupert Cole, Oliver Marsh, Tona Anzures, Elizabeth Jones, Jacob Ward, Julia Sanchez-Dorado, Erman Sozudogru, Raquel Velho, Toby Friend, Tom O'Donnell, Yin Chung Au, Yafeng Shan, Fr. Hugh Mackenzie, Jonathan Everett, Steph Ratcliffe, Sara Peres, and Matthew Paskins. Special thanks go to the late Bill MacLehose, whose untimely passing saddened many of us and left the department without a gifted teacher.

My current affiliation, the Faculty of History and the Research Center for the History of Modern World at Nankai University, has provided me with good office space and academic resources to complete this book. I especially thank the Nankai University Liberal Arts Development Fund for supporting my research and writing. I also thank the British Society for the History of Science for funding my research visits to the British Library and various conferences.

Many librarians, archivists, and museum curators assisted me throughout the research on which this book is based. While it is impossible to name them all, special thanks go to Sian Prosser, the librarian of the Royal Astronomical Society; Jane Harrison, the archivist of the Royal Institution; Sarah Hobbs, the archivist of the Manchester Archives (the Manchester Room and County Record Office); Alison Boyle, the curator of the Science Museum, London; Gloria Clifton and Richard Dunn, the curators of the National Maritime Museum, London; and Dinah Bott, the librarian of Priaulx Library, Guernsey. I also thank the staff at the Manuscripts Reading Room of the British Library, where I completed the transcription of Samuel James Arnold's *Ouranologia*. I thank Aileen Robinson for reminding me of the existence of this gem in the Lord Chamberlain's Plays collection.

The University of Pittsburgh Press has kindly provided indispensable assistance throughout the writing and publication of this book. As a novice author, I encountered inevitable delays, omissions, and mistakes. I am especially thankful to my editor, Abby McAllister, for her patient and professional support throughout the process. I also appreciate the comments and suggestions of two anonymous referees on my manuscript during the review process. Their advice has much improved the quality of this book.

Finally, my heartfelt gratitude goes to my father, mother, and little sister for their unwavering love and support. I am profoundly grateful to Christine Luk, for her love, inspiration, and encouragement over

ACKNOWLEDGMENTS

the years. If I have any courage to continue this academic journey, it all comes from my family's utmost care.

There is a saying in Chinese, "So many people to thank, and so we thank the heavens." I have been fortunate to encounter so many wonderful people during this endeavor. I regret that I cannot list all of them in these acknowledgments. To everyone, I say: Thank the heavens for this completed journey.

Note on Currency

In the nineteenth century, Great Britain did not use a decimal currency. The money was divided into three units: pounds (£), shillings (s.), and pence (d.). The guinea was a gold coin circulated before 1816, and its value was officially fixed at twenty-one shillings. After 1816, the term "guinea" was still common in colloquial usage and the charging of professional fees. The value of each unit is as follows:

    1 guinea = 1 pound and 1 shilling = 21 shillings (21s.)
    1 pound (£1) = 20 shillings (20s.)
    1 shilling (1s.) = 12 pence (12d.)

In this book, in addition to the above abbreviations, the notation £/s/d (with the abbreviations for shillings and pence omitted) is also used to represent monetary amounts. For example, £2/12/6 represents the amount of two pounds, twelve shillings, and sixpence.

COMMERCIAL AND SUBLIME

INTRODUCTION

A Sublime and Commercial Business

A *Chambers's Edinburgh Journal* article "The Central Sun" published in 1847 contains comments on the current situation of astronomy lecturing in Britain. Like many other articles in an all-inclusive magazine that provide knowledge about science, religion, history, and biography for common readerships, this one aims to introduce the topic of the sun's movement through space. Before getting into the main thesis, the anonymous journalist starts the article with a brief remark on the prevalence and popularity of astronomical lectures in recent decades. The journalist writes, "Lectures on astronomy have for many years been highly popular with a large portion of the public. . . . In the smaller provincial towns, the arrival of an itinerant lecturer, and the delivery of his 'course of three,' illustrated by an orrery, was an event productive of general satisfaction, and served to enliven one or two of the dreary weeks of winter." These astronomical lectures were presumably informative and entertaining, with extra amusements, as the journalist describes: "Something was generally added that largely excited the wonder of the auditors, who went away fully persuaded that they had learned the whole scheme and compass of astronomical science—for them it had no more secrets."[1]

The journalist's remark is a keen observation regarding the cultural phenomenon of popular astronomy lecturing at the dawn of the Victorian era. This was not an isolated testimony. Less than a decade earlier, around 1839, another anonymous journalist made a similar observation, even calling such enthusiasm for the knowledge of heavenly bodies "astronomical mania."[2] The journalist was reviewing the activities of eight lecturers, including the then renowned Dr. Dionysus Lardner, in London theaters during the previous Lent. Although Lent was an especially significant season for astronomical discourses, as we will see later in this book, Victorian contemporaries acknowledged that a lecture on astronomy was a "good card" at any season of the year.[3] According to the author of a bestselling treatise on astronomy, Ebenezer Henderson, popular astronomy lectures had been developed by several prominent figures, including James Ferguson, Benjamin Martin, and Adam Walker, during the eighteenth century through mechanical apparatuses and diagrams. Furthermore, lectures were rendered attractive in later years by the introduction of improved transparencies produced by optical devices. Henderson was also an astronomical lecturer and a fellow of the Royal Astronomical Society; he delivered his own course of twelve astronomical lectures in London toward the end of 1835 and was asked by friends to publish a treatise on the course.[4] The treatise was published in a third edition by 1848, marking his success in the area of scientific lecturing. His brief mention of those renowned names in this trade was not only an homage to his eighteenth-century predecessors but also an implicit affirmation that he was on the same enlightenment path to explain the principles of the universe for broad consumption.

Popular lectures on astronomy were a phenomenon in Britain throughout the nineteenth century. Lectures were presented in various places, and the speakers were not necessarily working astronomers. It is true that several celebrated elite astronomers, such as George Airy, John Herschel, and Robert Ball, made significant efforts to popularize astronomical knowledge. Their cases have been thoroughly discussed in scholarly works.[5] Nevertheless, many more popularizers of astronomy were not among the scientific elite. Historians have paid scant attention to these institutional and private entrepreneurs.[6] Some astronomical shows were linked with, but not limited to, Lent, and featured transparent orreries as visual aids. Private lecturers' contributions to popular astronomy remain obscure despite the widespread popularity they enjoyed at the time.

INTRODUCTION

This book explores the wide spectrum of popular astronomy lecturing in Britain during the Regency and early Victorian eras. Although I trace the pioneers of the astronomical lecturing trade to the 1780s and earlier, I focus on the period between 1820 and 1860, the heyday of private astronomy lecturing trade and theater-based Lenten lectures on astronomy. They were on the wane after the mid-nineteenth century. This period also witnessed the rapid growth of scientific institutions, ranging from specialist learned societies and literary and philosophical institutions in cities or provincial towns, to mechanics' institutes aiming at education of the working classes. Many of these institutions offered public scientific lectures and became hubs of local intellectual life. Despite this trend of institutionalization in science, private entrepreneurs occupied a notable place in the popular astronomy lecturing trade and continued thriving until the 1860s. They were not inferior to their institutional competitors in terms of popularity with and influence over audiences. I compare the activities of private entrepreneurs with those of institutional men of science, such as the discourses delivered by Airy at the Royal Institution of Great Britain.

There are several good reasons to focus on private lecturers of astronomy outside the scientific elite. Besides the fact that they have received less attention from previous scholars, often, private entrepreneurs were not career practitioners of science. Their agendas and styles of lecturing differed from those of institutional men of science with which we are familiar. Historical studies of "popular science" and its agents have been a growing field in recent decades. Researchers view popular science variously as the popularization of scientific knowledge or as the representations of science in popular culture.[7] Bernard Lightman distinguishes the popularizer from the practitioner: the former focus on conveying scientific knowledge, whereas the latter engage in original research such as conducting experiments and analyzing natural worlds.[8] This distinction is not meant to be rigid: some practitioners of science—for example, Michael Faraday and John Tyndall—were also keen on popularization. Lightman, however, emphasizes a group of "nonpractitioner" popularizers, who offered sensational science to the public through writing or lecturing, but whose agendas often disagreed with the scientific elite's. The case studies presented in this book will show that many private entrepreneurs of astronomy operated independent lecturing businesses without affiliation to scientific establishments. In contrast, institutional lecturers were usually associated with or employed by literary and scientific learned institutions. To analyze astronomy lecturing in the context of the popularizer's affiliation and the sites where they performed can help historians to better understand the complexity of the social and cultural milieu of science in the nineteenth century.

Here I must again defend the historical studies of popular science. Historians of science used to dismiss such subjects as irrelevant to the history of science "proper." In addition to historians, today's scientists often have pejorative attitudes toward popular science, believing that its practice accrues no benefits to their professional career. This was one reason that the "public understanding of science" movement emerged in the United Kingdom after 1985—not only to enhance the public's scientific literacy but also to encourage scientists to engage in communicating science to lay audiences.[9] In describing how nineteenth-century scenes of opportunities to buy and sell goods and services are related to the display of knowledge, James Secord uses the term "commercial science" for these various activities combined, rather than the easily dismissive catchall "popular science."[10] Nevertheless, popular science mattered throughout the development of institutionalized science, for the relationship between the two was dynamically changing and never fixed. The definitions and boundaries between the two were porous for most of the nineteenth century. Instead of a one-way process of knowledge transfer from elite science to the lay public, popular science worked as a "heterogeneous network of cultural exchanges and feedback loops between different social groups." Expert and nonexpert knowledge making often overlapped, forming two-way, interactive flows.[11] In astronomy, recent studies also reveal that nineteenth-century astronomers use the popular press as a channel to influence elite scientific practice and to promote public interest.[12] As Joshua Nall put it, "astronomers were journalists and editors too, eliding practice with communication in consequential ways."[13] Therefore, popular astronomy also contributed to the shaping of a new institutionalized astronomy.

As Ebenezer Henderson indicated in his 1843 treatise, thanks to the introduction of improved transparencies and optical technologies, popular astronomy lecturing underwent a welcome transformation during his time. Henderson's remark was an understatement; in fact,

INTRODUCTION

this transformation was not only limited to the adoption of new visual aids but also involved a broader revolution in lecturing space and format. I argue that popular astronomy lecturing underwent a "theatrical turn" before the 1820s. This transformation related to the use of the theater, both the physical space and the showmanship employed by professional actors. The physical space of a site was a significant factor in shaping the format and style of lecturing as well as the audience for it. Astronomical lectures performed in some of London's famous West End theaters best exemplified this spatial effect. Some popularizers moved astronomical lectures into theaters and adopted theatrical devices in their performances. These lecturers developed numerous onstage devices for achieving scenic effects, and their performances emphasized both sensational amusement and scientific instruction. This trend began in the late eighteenth century and continued developing throughout the early nineteenth century. It benefited from the invention of the transparent orrery, a type of apparatus designed as an onstage visual aid for a large audience.

The theatrical turn of popular astronomy lecturing shows the diversity of the narrative and practice of science in the mass culture of the industrial age. Whether they were private entrepreneurs or institutional employers, popularizers of science competed to attract the largest audience. They had to consider the interests and tastes of the public. Astronomical lecturing had become such a potentially profitable business that many competitors entered this marketplace. It was precisely a part of the vast network of what Secord calls commercial science. On the other hand, using theatrical facilities and apparatuses to create spectacular displays of the celestial system not only enhanced the quality of amusement but also offered spiritual inspiration. The language and visualization often presented in astronomical lectures made a science that was traditionally linked with the glory of God's handiwork even more sublime. "Commercial" and "sublime" are therefore the two keywords that best describe the character of British popular astronomy lecturing in the first half of the nineteenth century.

THE SUBLIME SCIENCE

Astronomy was the epitome of the sublime for many nineteenth-century contemporaries. When reporting astronomical lectures in newspapers and magazines, Victorian journalists often referred to the studies of heavenly bodies as "the sublime science."[14] Thomas Milner in his extensively illustrated bestseller on astronomy and geology, *The Gallery of Nature* (1846), also described astronomy as "now the most perfect of all the sciences" and "the most sublime and ancient."[15] Similar descriptions can be traced to eighteenth-century authors such as James Ferguson in his opening remark in his classic *Astronomy Explained upon Sir Isaac Newton's Principles* (1756): "Of all the sciences cultivated by mankind, astronomy is acknowledged to be, and undoubtedly is, the most sublime."[16] These examples reveal that people of that time valued astronomy and elevated its study to a special status.

The common meaning of the term "sublime" is twofold according to the *Oxford English Dictionary*: the first is relevant to rational or spiritual achievement, a premier quality belonging to or designating "the highest sphere of thought, existence, or human activity; intellectually or spiritually elevated."[17] Astronomy had been regarded as the highest achievement of humanity, for the science attempts to decipher the code of the universe by human reason. The dissemination of scientific knowledge also indicates a path to reason. Humans benefited from the advantages of science and no longer linked natural phenomena with irrational superstition. Such progressive enlightenment narratives were common in scientific publications at the time, and the commentators included some of the most prestigious men of science. William Whewell, for example, asserted the benefits of astronomy for learning the character of the government of the world: "In considering the universe . . . as a collection of *laws*, astronomy, the science which teaches us the laws of the motions of the heavenly bodies, possesses some advantages."[18] John Herschel claimed that no science other than astronomy required the highest preparation of minds and intellectual liberality to see through "superficial and vulgar" observation.[19] Another meaning of "sublime" is in the scope of emotion: a sense of emotional uplift that "fills the mind with a sense of overwhelming grandeur of irresistible power," and such emotion "inspires awe, great reverence, or other high emotion, by reason of its beauty, vastness, or grandeur."[20] The immensity of the universe and the vast dimensions of celestial bodies are ideal for inspiring the feeling of the sublime. Both uses of the adjective, whether indicating intellectual or emotional uplift, had emerged and were commonly applied in the English language by the eighteenth century.

INTRODUCTION

The concept of "sublime" had been much discussed in deeper philosophical or theoretical expositions. Edmund Burke's famous treatise *A Philosophical Enquiry into the Origin of Our Ideas of the Sublime and Beautiful* (1757), for example, represented attempts to distinguish the senses of the sublime and beauty. Unlike the sense of beauty, caused by exquisite and pleasant things, the source of the sublime according to Burke is the terrors of pain, danger, and death. The sublime can produce strong emotion, hence passion. Burke regards the emotions of astonishment, admiration, and reverence as the effects of the sublime. He also links the sublime with other visual aspects such as vastness, infinity, and darkness. Burke's exposition does not directly involve astronomy, but it is noteworthy for the connections it makes between the sublime and passion. Burke's work is merely one of many examples of intellectuals pondering the concept of sublime in and before the nineteenth century.[21] Nevertheless, as Jan Golinski reminds us, the term "sublime" was often deployed casually rather than systematically in contemporary writings, and the nuances of its usage in learned debate were not necessarily connected to popular astronomy.[22] Newspaper journalists and authors of popular astronomical publications, after all, were usually not serious theorists or philosophers. Their accounts did not necessarily carry profound philosophical reflections.

From Burke's treatise, it is not difficult to imagine that the notion of sublime could be highly spiritual and connected with religious belief. Since feelings of the sublime can be linked with strong emotions, reverence for magnificent power, and a sense of elevation, such emotion is commonly evoked through religious experience and rhetoric. This book discusses the rich religious elements in popular astronomy lectures of the nineteenth century. Lecturers often applied rhetorical devices from then popular natural theology, including the idea of sublimity: the universe is an unparalleled source of inspiration for feelings of awe and wonder. People discerned the existence and divine wisdom of a benevolent Creator through the majesty and order of the universe, God's handiwork. Such a religious trend accorded with many popular science works at the time, as studies of the Bridgewater Treatises and evangelical publications have shown.[23] Theological reflections were not necessarily high on the agenda of all astronomical lectures, but religious and relevant moral inspiration had become a good way for lecturers to promote popular astronomy.

INTRODUCTION

The theatrical turn of popular astronomy lecturing was critical to solidifying astronomy as the sublime science in popular culture. Adam Walker and his sons, who allegedly invented the transparent orrery in the late eighteenth century, initiated the theatrical turn of the trade and made astronomy a star in show business. The Walkers pleased the audience's minds, eyes, and ears, as they incorporated scenic displays and euphonious music into their lectures. Michael Faraday once commented on the lecture of William Walker, Adam Walker's eldest son, that he "has shewn in the most splendid and sublime manner that Astronomy may be illustrated."[24] Visual and acoustic technologies introduced into the performance helped enhance the sense of the sublime. These techniques shaped astronomy as the sublime science, in which awe and wonder of the universe had become a universal theme.

Science in the Marketplace

The spread and popularity of astronomical lecturing also has to be examined in the context of both the scientific and consumer cultures in the industrial age. The growing power and pervasiveness of science in daily life had become such a powerful phenomenon that nineteenth-century contemporaries found it impossible to overlook. "Science is no longer a lifeless abstraction floating above the heads of the multitude," as an author described it in 1852, "it has descended to earth. It mingles with men. It penetrates our mines. It enters our workshops. It speeds along with the iron courser of the rail."[25] Science, now a fashionable conversation piece, drew the general public's attention and fascination. Science also had many practical uses. The Whig politician Henry Brougham asserted that men could gain positive advantage in worldly wealth and comforts by increasing their stock of information.[26] As an avid reformer of scientific education, Brougham founded the Society for the Diffusion of Useful Knowledge (SDUK) in 1826. A series of SDUK publications was launched during the next two decades under the supervision of the publisher Charles Knight, including the *Penny Magazine* and the *Penny Cyclopaedia*. The founding of the Royal Institution of Great Britain at the turn of the nineteenth century also embraced the vision of popularizing useful science. The proposal for this new establishment sought to launch an institution for "diffusing the Knowledge, and facilitating the general Introduction, of Useful Mechanical Inventions and Improvements; and for teaching...

the application of Science to the common Purposes of Life."[27] Within the next few decades, the Royal Institution consolidated the image of science as a rational and practical agent of improvement and change.

The milieu of science was undergoing a tremendous transformation during the same period. Historians of science concur that the long nineteenth century was a pivotal age during which many characteristics of modern science were taking shape.[28] The younger generation of scientific intellectuals such as Charles Babbage claimed that British science was declining; the British Association for the Advancement of Science (BAAS) was set up to avert the decline and to reform the sluggishness of the Royal Society. Whewell coined the term "scientist" at the meeting of the BAAS in 1833 and in a subsequent book review argued that the term was, like "artist," appropriate for those pursuing the study of nature. Initially, Whewell's use of the term was not intended to be serious, and not until the twentieth century did the practitioners of science embrace it.[29] A career in science was previously unprecedented, but the process that transformed science into a profession was also beginning.

The advancement of technology and industry also revolutionized the communication of scientific knowledge. Steam-powered printing machines made inexpensive mass printing possible. Railway, telegraphy, and postal systems provided faster, more efficient means of transportation and transmission. The spread of religious tracts, Sunday schools, and secular education also cultivated greater literacy among the middle and working classes. All these factors changed the businesses and culture of publishing, distribution, and reading.[30] A new type of readers emerged and numerous low-priced scientific publications for ordinary readers were sprouting by the mid-nineteenth century. This wave of literature included Brougham's and Knight's ambitious *Penny Magazine* (1832–45) and other SDUK publications (e.g., *Penny Cyclopaedia*, 1833–43). The periodical *Chambers's Edinburgh Journal* (1832–1956) coedited by the Scottish publishers William and Robert Chambers was another representative example among the army of affordable publications.[31] Robert Chambers anonymously wrote another more scandalous work, *Vestiges of the Natural History of Creation* (1844), to promote the controversial theory of evolution, which caused an immediate and lasting sensation. In addition to secular reformers, Christian denominations were also keen to popularize scientific knowledge, and they were a force to be reckoned with in the marketplace of popular science publishing. The representative works endorsed by religious authorities included the Bridgewater Treatises (1833–36), based on natural theology and written by several celebrated scientific savants, and the "Monthly Series" (1845–55) published by the evangelical Religious Tract Society (RTS).[32]

When the transformation of science led to the coining of the term "scientist," the expression "popular science" also came into use. Jonathan Topham indicates that publications before 1820 seldom appeared under the designation "popular science," yet the phrase rapidly came into regular usage after that. The new genre of popular publications was born of a social change. It appealed to a broader audience: a new class of readers covering not only the Enlightenment bourgeois public but also the working classes. The popular science genre had multiple meanings: it could refer to a text that made scientific knowledge more accessible to a general reader or to publications that were mass-produced and more affordable. Both meanings indicate the burgeoning mass culture of the early nineteenth century. Therefore, some historians consider the nineteenth century as the dawn of popular science.[33] The rise of popular science, in Topham's words, reflected a "diversification of reader audiences," a movement to accompany the trend of specialized and disciplinary science, and was "loaded with consciousness of the new social order."[34]

In addition to print culture, lecturing was another important channel for diffusing scientific knowledge. Public lectures on natural philosophy had been presented in Britain since the early eighteenth century.[35] Their business mode became the precursor of many scientific lectures including astronomical ones (see chapter 1). During the first half of the nineteenth century, public lectures not only grew in audience scale but also were increasingly conducted by the fast growth of scientific institutions, such as the Royal Institution and the London Mechanics' Institution in the British capital. Private or itinerant lecturers who operated their own businesses and traveled around provincial towns remained active alongside their counterparts employed by scientific institutions.

Shows, exhibitions, and different types of entertainment displays other far-reaching forms of media alongside print culture and public lectures. In big cities like London, this form of media in particular

INTRODUCTION

thrived and dazzled. Various exhibitions, shows, and spectacles were staged in the metropolis to appeal to spectators' intellectual curiosity and amusement. The Adelaide Gallery (1832–52) and the Royal Polytechnic Institution (1838–81) were two renowned examples among the splendid array of metropolitan galleries. The British were proud of their love of shows, as a letter to the newspaper the *Examiner* remarked: "Well might the great Napoleon say, we trafficked in every thing; but he was little aware that to 'a nation of shopkeepers,' he might have added, of show-keepers."[36] Among numerous bizarre shows in the metropolis, exhibitions and spectacles displaying scientific curiosity or mechanical ingenuity occupied a distinct part of this marketplace. A pioneering classic by Richard Altick, *The Shows of London* (1978), documents a wide range of entertainments and exhibitions in London during the Georgian and early Victorian periods until the Great Exhibition of 1851. Many of the cases Altick investigates, such as the exhibitions of automatons and other mechanical inventions in the West End, involve applications or disseminations of scientific novelties.[37] Following his work, scholars have also paid increasing attention to the function of performativity in scientific practice as a means of attracting audiences or reinforcing cultural authority.[38]

These numerous enterprises, as Aileen Fyfe and Bernard Lightman indicate, were a part of the economy involving the display of knowledge in the marketplace of science.[39] Scientific businesses took place in various venues ranging from learned institutions, museums, botanical gardens, shops, and theaters to exhibitions and spectacles. Practitioners of science could earn their livelihoods from a variety of activities including authorship, editing, lecturing, curatorship, instrument making, show-running, and so on. A metropolis like London had abundant opportunities for entrepreneurs who wanted to make money or young people who aspired to build a reputation in scientific circles. The clichéd story of Michael Faraday rising from a post as laboratory assistant to become a star professor at the Royal Institution is one of many examples. Among Faraday's contemporaries were many writers who composed scientific pieces for periodicals and monographs, as well as performers who demonstrated experiments with current and sparks in places like the Royal Polytechnic Institution and the Colosseum at Regent's Park. Contemporaries and later researchers often dubbed these myriad activities "popular science." This term is, however, not without problem as many scholars have indicated. Recall, for example, Secord's criticism of the phrase "popular science" as a dismissive catchall that could easily mislead by confusing its historical significance with today's perceptions. Thus, Secord suggests using the term "commercial science" as an alternative.[40] Nevertheless, like many other alternatives, this expression cannot cover the complex and diverse facets of relevant engagement in history, so it is not necessary to abandon the term "popular science," and some researchers retain it.[41]

A Shared and Contested Arena

So far, I have set the scene for the theatrical turn of popular astronomy. The tradition of itinerant public lectures on natural philosophy, together with the fashion of stage astronomy, formed the cornerstone of popular astronomy lecturing at the turn of the nineteenth century. The industrial boom and economic prosperity of Britain during this period prepared a growing market for readers and audiences; social and political upheavals influenced the taste for and representation of popular science. The prevalence and popularity of astronomy lecturing in the nineteenth century was no isolated development, and it did not appear from nowhere. It was an extension of the previous century's legacy as well as a reflection of the noticeable social change in the early nineteenth century. Like other contemporary spectacles of chemistry, electricity, and geology, which have been explored by other scholars, public displays of astronomy demonstrate the zeitgeist and the transformation of society.

Many scholars have indicated that nineteenth-century science was a contested space in which rival notions of how and by whom legitimate knowledge should be constructed competed and were promulgated.[42] This perspective suggests a decentering approach to the sites, actors, and practices of science; as Iwan Morus remarks, the locus of scientific authority was diverse—"both everywhere and nowhere."[43] This approach also echoes Lightman's attempt to reconstruct a distinctive place for nonpractitioner popularizers in the topography of nineteenth-century British science. The agendas of these popularizers were usually at odds with institutional elites of science.[44] Popular astronomy lecturing in the nineteenth century embodied such a contested arena, where different classes of popularizers coexisted and competed, performing with potentially opposed agendas in various venues.

INTRODUCTION

"Commercial" and "sublime," I argue, are the two aspects that best characterize British popular astronomy lecturing in the nineteenth century. As we have seen, previous scholars have approached the popular products and phenomena of science from the perspective of mass consumer culture, whether using the phrases "science in the marketplace" or "commercial science" to designate the hotchpotch of paid pursuits of science in the period. It is true that not all popularizers of science intended to make a profit—some were motivated by religious, political, or philanthropic causes and not concerned with money in the first place. Nevertheless, the approach that treats science in the context of mass consumer culture is valuable for highlighting the economic factor and everyday operations in the practice of popularizers.[45] The concept "commercial" can thus serve as an analogy between popular astronomy and profitmaking, and as a way to interpret the role of audiences as consumers of science. It also underscores the competition among astronomy lecturers to gain audiences, develop innovative apparatuses, and protect trade secrets. In this book I will show the rivalry among metropolitan lecturers during Lent; improved apparatuses, showmanship, and advertisements for marketing; the itinerant lecturing circuit between the metropolis and the provinces, and the audiences that followed a fashion, a cause, or a utility. These activities involved competition for profits and commercial practices of buying and selling. Even "nonprofit" institutional establishments, such as the Royal Institution, were more or less involved in some commercially oriented practices.

The adjective "sublime," in contrast, was an original description extensively used by nineteenth-century contemporaries rather than an invented term reconstructed by modern scholars. Nineteenth-century popular science was filled with emotional appeal. Authors and lecturers loved to arouse feelings of awe, wonder, and pleasure by using passionate language in their narratives. Scientific issues were also associated with other spiritual concerns like morals and Christianity. Discourse about the heavenly bodies was usually relevant to earthly orders, too. As Golinski suggests, the use of the term "sublime" might allow lecturers to introduce some potentially subversive ideas, such as the plurality of worlds, under the cover of religious or quasi-religious sentiments.[46] Popular astronomy lecturing made extensive use of different means, from the rhetoric of natural theology to the application of visual and

INTRODUCTION

acoustic technologies, to render the sublimity of astronomical science. The "sublime" aspect also reflects the richness of religious elements in nineteenth-century popular astronomy. The spiritual perspective of "sublime" is therefore a good supplement to the material perspective of the commercial culture.

Many nineteenth-century practitioners of popular astronomy, as well as their activities, exemplified the commercial and sublime features. The appendix provides a full list of the lecturers discussed or mentioned in this book. The book focuses on individuals who were active in London and regularly operated lecturing businesses between 1820 and 1860. Some prominent names among the popularizers, such as the Walker family, have frequently been discussed in previous scholarly works. Some lecturers, such as C. H. Adams and John Wallis, have been mentioned in literature before, but few details of their activities or biographical information are known. I supplement the latest biographical findings of these two individuals in this book. Two figures are recognized for other occupations yet their involvement in popular astronomy lecturing is not widely known: George Bartley is recognized for his career in the theater, and George Henry Bachhoffner is discussed by historians of science for his demonstrations of electricity at the Royal Polytechnic Institution.[47] Several notable men of science who delivered public lectures on astronomy, such as George Airy, John Pond, John Pringle Nichol, and Baden Powell, are also covered in this book. Nevertheless, my research is not intended as a complete survey or biographical account of particular astronomical lecturers.

The structure of this book is thematic rather than biographical or chronological. Each chapter discusses one theme related to a specific area of astronomy lecturing. Chapter 1 considers the pioneers of popular astronomy lecturing in the late eighteenth century, whose discourse and apparatuses profoundly influenced subsequent popularizers. Chapter 2 explores the geography of popular astronomy lecturing, to investigate the different venues in metropolitan or provincial regions where popular astronomy took place. In other words, the chapter addresses the question of "where." Chapter 3 deals with the affiliations of lecturers. By analyzing the relations between lecturers and institutions, the identity and place of a lecturer within the web of scientific practitioners is revealed—that is, the issue of "who." Chapter 4 concerns the subjects that were included in the curricula of pop-

INTRODUCTION

ular astronomy. Some recurring subjects contained not only widely accepted Newtonian science but also controversial issues like the nebular hypothesis and the plurality of worlds. We also find that scientific novelty and religious sentiment were strong attractions for the contemporary audience. This chapter relates to the questions of "what" and "how," as does chapter 5, which explores the use, popularity, and constraints of the transparent orrery and lantern slides—the two major types of apparatus used in nineteenth-century popular astronomy lecturing. The mechanism of the transparent orrery has remained a matter of debate until recently. I will extrapolate its possible nature from the latest available sources. This large-stage apparatus certainly had a prominent place in nineteenth-century popular astronomy, yet very little literature and almost no physical remnants have survived. Application of the transparent orrery in stage astronomy influenced other visual technologies such as the magic lantern. Finally, chapter 6 discusses the audiences for popular astronomy, in which a few accounts of contemporaries are presented as specimens reflecting the fashion of astronomy lecturing, the responses from spectators, and the conflicts between various stances concerning science.

CHAPTER 1

PIONEERS

Rome was not built in a day, and neither was popular astronomy lecturing. Astronomical lectures in the early nineteenth century did not pop up out of nowhere. They evolved from public lectures on natural philosophy in the prior century. One can easily find traces of experimental philosophy lectures in the lecturing trade of popular astronomy before the mid-nineteenth century, particularly in the patterns of lecturers' careers, discourses, subjects, apparatuses, and itineraries.

In a pioneering article dedicated to the historical significance of popular astronomy, Ian Inkster looks at public lectures in Britain from the mid-eighteenth century to the early Victorian era and argues that astronomy was peculiarly suited as a "popular" science during the period. Inkster indicates that a variety of appeals helped astronomy endure in public life. These appeals included the lectures' substance and implications, which offered "alternative or additional attractions in the areas of morality, religion and political economy," rather than utility of the science itself.[1] He also traces the construction of this social (or cultural) role of astronomy to around 1750, when "the implications of Newtonian synthesis were worked out, digested and elaborated," and thus he examines popular astronomy since then.[2] Inkster is rightly aware that the prosperity of popular astronomy before the mid-nineteenth century was inseparable from the development of Newtonian science.

Therefore, before we examine popular astronomy lectures in the nineteenth century, it is necessary to understand what was going on among those predecessors in the eighteenth century. The trend of science to become a public culture in Britain was the result of the experimental philosophy promoted by the Royal Society. It was also the dawn of the time when science reached a wider audience and took root in the public sphere of the Enlightenment.

Public Lecturing on Natural Philosophy in the Eighteenth Century

Public lecturing on natural philosophy was a new and burgeoning trade in eighteenth-century Britain. It was a cooperative endeavor promoted by the Royal Society's experimental methods, Newtonian science, and the flourishing scientific instrument making trade in England. The Royal Society (the Society) was famed for its advocacy of experimental philosophy, of which Robert Boyle and his assistant Robert Hooke were among the representative champions in the Society's early history. Hooke was appointed as the Society's first curator of experiments in 1662, a post demanding the regular demonstration of experiments before its members. The Society recognized the utility of public demonstration by the beginning of the eighteenth century: the practice transformed from its original purpose of producing knowledge before a specific group of "gentleman" witnesses to a wider audience for spreading knowledge.[3] In addition, the public demonstration of experiments was also a remedy for the esoteric mathematics required in Newtonian science. Few of Newton's contemporaries had mathematical skills comparable to those of Newton himself. By demonstrating experiments through general or purposely made apparatuses, Newton's disciples could popularize Newtonian science in a more accessible way.[4]

Thus, it is no surprise that several pioneers of public lectures on natural philosophy were fellows of the Royal Society. John Harris was among the earliest to deliver public mathematical lectures in London. His talks, delivered between 1698 and 1707, contained topics of mechanics including the use of globes and various mechanical devices. Harris's lectures heralded what was to come. The earliest public lecture titled "Natural Philosophy" was delivered by James Hodgson in 1705. Hodgson also focused on astronomy, declaring in the advertisement for the lecture that it was for "the advancement of Natural Philosophy and Astronomy" and "the benefit of all such Curious and Inquisitive Gentlemen as are willing to lay the best and surest Foundation for all useful Knowledge." His lecture provided instruction in apparatuses such as "Engines for Rarafying and Condensing Air," microscopes, telescopes, prisms, barometers, thermometers, and "Utensils proper for Hydrostatical Experiments, in order to prove the Weight and Elasticity of the Air."[5] These apparatuses were produced by the Royal Society's chief experimentalist Francis Hauksbee (the elder), who was a skillful instrument maker and Isaac Newton's laboratory assistant. Hodgson and Hauksbee collaborated to conduct the lecture and published a pamphlet of the lecture's syllabus afterward.[6]

Another early celebrity of public natural philosophy lectures was John Theophilus Desaguliers (1683–1744), who also had a profound influence on the development of astronomical lectures. Desaguliers succeeded Hauksbee in his post at the Royal Society as the curator of experiments in 1716. A descendant of émigré French Huguenots who received university education at Oxford, Desaguliers was fluent in English, French, and Latin, and this language and academic literacy proved to be a great asset to his scientific career. Through his writing, translation, and lectures on Newtonian philosophy, Desaguliers maintained contacts with continental scholars and contributed to the international spread and consolidation of Newtonian science. His commercial lectures on natural philosophy outside the Royal Society were successful and influential, for many younger experimental philosophy lecturers were his students.[7] Desaguliers was also the earliest natural philosopher to recognize the educational value of the orrery, which was then still a novel invention, in astronomy and to incorporate it into lectures. "Orrery" was an umbrella term loosely used to describe mechanical models of the solar system of various degrees of complexity and types of mechanisms. Orreries were invented in the early eighteenth century and evolved through the development of public lecturing on natural philosophy. They were soon used by lecturers as a visual aid to demonstrate the motions of celestial bodies and the causes of relevant phenomena. In 1717 Desaguliers described the Planetary Machine made by George Graham as capable of showing "the Motion of the Earth and Moon about the Sun, and theirs and the Sun's Motion about their own Axes, as also the Inclination of the Earth's axis, always the same."[8] He later built a planetarium with his own design and described this apparatus in the treatise *A Course of Experimental Philosophy* (1734).[9]

Those early public lectures on natural philosophy conducted by

Hodgson, Hauksbee, and Desaguliers already exhibited some characteristics that were shared by subsequent followers during the eighteenth century. The first feature was the commercial nature of these lectures. Experimental philosophy and all related things, from lectures and books to instruments, had been commoditized in Newton's time, and its promotion grew with the emerging market of newspapers and periodicals by the mid-eighteenth century. "Science was a commodity, a thing to be purchased, displayed, consulted, and contemplated," as Jeffrey Wigelsworth remarks, according to his investigation of advertisements for scientific commodities in contemporary newspapers.[10] Although early lecturers were usually fellows of the Royal Society, they conducted lecturing activities outside the institution and managed their own businesses just like merchants marketing a brand.[11] The courses charged admission. Attendees usually needed to pay 2 guineas—one at the time of subscription and the other after the course had begun. The lectures usually took place in a coffeehouse or at the lecturer's residence. Other auxiliary products like textbooks or syllabi were also sold by the lecturers or partner booksellers.

Another characteristic of public lectures on natural philosophy was the significant instrument orientation of the curricula. Exhibitions and demonstrations of a variety of apparatuses were often incorporated into the lectures as core subjects. Some of the apparatuses were general experimental or mathematical instruments like air pumps, globes, and telescopes. Others were purposely made devices to show the effects of natural laws behind mechanical principles, such as the model "Maximum Machine," originally designed by Desaguliers for measuring the maximum of a man's power to raise water.[12] A trademark feature of experimental philosophy lectures, as shown in Hodgson's advertisement, was the practice of enumerating the apparatuses that would be displayed in the lecture. Later lecturers continued this tradition. For example, William Griffis advertised his lectures with "Celebrated Philosophic Apparatus" in Derby in 1743, which included a vast variety of instruments such as "a curious Air Gun, a Planetarium, a Cometarium, a Ptolemaic Sphere, and an improv'd Orrery, Tellescopes and Microscopes of all sorts."[13] Many experimental philosophy lecturers either collaborated with an instrument maker, as in the joint venture between Hodgson and Hauksbee, or they were themselves manufacturers of instruments.

The success of early experimental philosophy lecturers drew followers, and those new to the market were soon engaged in business rivalries. The Royal Society did not have a monopoly on public lecturing, as later lecturers were not necessarily its fellows. Nevertheless, having a fellow title or the patronage of a noble aristocrat still boosted the lecturer's reputation. Alan Morton and Jane Wess have surveyed the advertisements of public lectures on natural philosophy in the London newspaper *Daily Advertiser* between 1745 and 1770. More than twelve people advertised their lectures in the paper during this period, including some of the most preeminent figures in the trade, like Desaguliers's nephew and student Stephen Demainbray, the self-taught Scottish inventor James Ferguson (1710–76), and the English entrepreneur Benjamin Martin (bap. 1705, d. 1782).[14] The latter two were particularly crucial to popular astronomy and will be discussed in chapters 4 and 5. In some peak seasons such as November and December 1757, 1758, and 1760, four different advertisements simultaneously promoted experimental philosophy lectures.[15] From the number of advertisements and their clashes during a short period, we can easily see the growth of the lecturing trade and competition in the market.

Philosophy lecturers' activities were not confined to London. Although many lecturers and instrument makers were based in the metropolis, their businesses often expanded across the country. While some provincial lecturers were active in a particular region, other itinerant lecturers traveled between towns. Griffis, for example, began by touring around Midlands towns such as Derby, Birmingham, and Wolverhampton, but by 1757, he had expanded his businesses to West Country towns including Bath, Bristol, and Salisbury.[16] Demainbray started his lecturing career in Edinburgh in 1749 and then moved to Newcastle and Dublin. He even crossed the English Channel to lecture in France, following in the footsteps of his mentor Desaguliers. By 1755 Demainbray had eventually settled in London. His philosophical instruments, combined with King George III's collection, were later acquired by King's College London and are now on display at the Science Museum in London.[17]

To attract audiences and promote scientific reputations, lecturers often published accounts of their lectures in addition to advertisements in the newspapers. These works could be a concise syllabus, a pamphlet, or even an elaborate treatise. Two titles by the lecturer John Arden and his son James Arden best exemplify this promotional strategy. The first was an advertising pamphlet, *A Short Account of a Course of Natural and Experimental Philosophy* (1772), which contained only eight pages and simply listed the subjects of each lecture. However, it summarized the admission charge (1 guinea), the rules of subscription, and the benefits of learning experimental philosophy. John Arden, a relatively obscure figure, self-styled as a "Teacher of Experimental Philosophy" at Beverley, is best known by later biographers for teaching young Mary Wollstonecraft.[18] The subjects in Arden's course included natural philosophy in general, mechanics, astronomy, geography, hydrostatics, pneumatics, and optics. The lecturer also promised that the experiments would be rendered as plain and intelligible as possible, and the apparatuses he demonstrated in the course were "elegantly finished with [the] *latest Improvements*."[19]

Arden's son James was credited with the other work, *Analysis of Mr. Arden's Course of Lectures on Natural and Experimental Philosophy* (1774). This little book, containing sixty-four pages, was a synopsis of the lectures. In the short preface titled "Advertisement," James Arden claimed: "Having frequently heard the Subscribers to this Course request the Publication of some such Treatises as this, to serve by way of Memorandum or Pocket-Companion, for those Parts which would be most likely to escape the Memory; and at the same Time knowing that my Father was too much engaged to Business, to comply with those Requests; induced me to draw up this Analysis; which will I hope, in some Measure, answer the End proposed."[20] Clearly, based on this short preface, *Analysis of Mr. Arden's Course* served as a kind of lecture note or even a "souvenir" for the audience of the course. It was not unusual for a father and sons to jointly operate a family business of lecturing or instrument making at the time. The arrangement of authorship was therefore done on purpose, for it could promote the lectures of Arden the senior, while also providing Arden the junior with some visibility. While *A Short Account of a Course* was pure advertisement, the treatise *Analysis of Mr. Arden's Course* was the merchandise following the course. The price—1 shilling and 6 pence—was printed on the title page. This piece of commodity was ready to be sold to eager natural philosophy students, whether they were subscribers to Arden's course or readers from afar unable to attend in person.

A Stellar Subject in Polite Science Culture

Astronomy had been a key and conventional subject in popular science ever since public lecturing of natural philosophy developed in the eighteenth century. It was considered by Enlightenment thinkers and authors as the ideal rational science—a science that best demonstrates how humans had attained a rational understanding of the world, especially an orderly (and correct) cosmology built upon Newtonian principles. The laws behind the revolution of celestial bodies had been deciphered by the power of reason. Fears of strange phenomena in the sky like eclipses and comets were dispelled by the comprehension of astronomical knowledge. As John Bonnycastle (a teacher and prolific writer of science in the late eighteenth and early nineteenth centuries) claimed, "Enlightened as we are at present," the old superstition "was once the general language of mankind; and it is Astronomy alone that has delivered us from these evils."[21] In addition to rational understanding, especially from a Christian perspective, astronomy was also spiritual. Unraveling the laws of the natural world is to know the will and wisdom of the Creator behind the universe. William Whewell, the polymath Cambridge don and Anglican priest, designated astronomy as the "queen of sciences" and the "only perfect science," as it had reached an elevated state of flourishing maturity. To Whewell, astronomy was the only branch of human knowledge in which "particulars are subjugated to generals, effects to causes." Long observation of the past could receive a prophecy through reason, and thus humans were able "fully and clearly to interpret Nature's oracles."[22] The above causes gave astronomy a particularly morally uplifting aspect, freeing people from superstitious belief while turning their minds to the powers of the Creator.

Another factor that contributed to the spread of astronomy as a popular subject was the polite culture of eighteenth-century British society. Appropriate etiquette and politeness in social situations was fashionable among the contemporary upper and middle classes. Courtesy was linked to material wealth as well as civility. The economic growth and the coming of the industrial age created a growing middle class in Britain, which loosely included anyone who could sustain a proper financial status such as businesspeople, industrialists, professionals, shopkeepers, and craftspeople. Members of the middle class emulated gentility in their material wealth. The pursuit of politeness was associated with the possession of goods, which marked a distinction from the lower social classes. In a broader sense, politeness was associated not only with material belongings but also with intellectual and aesthetic tastes. Society would judge adherence to a certain code of conduct as civilized; this included engaging in cultivated pursuits.[23] By participating in intellectual activities such as public lectures and conversations, and collecting genteel and luxury commodities, middle-class people attempted to create a social image for themselves that was distinct from the common "vulgar" sort.

Polite pursuits were also associated with booming businesses of natural philosophy. Science itself became a part of the polite culture in the public and domestic spheres, a phenomenon that Alice Walters calls "polite science."[24] To eighteenth-century upper- and middle classes, studying subjects of natural philosophy, astronomy, and geography and having a proper sense of the sciences were necessary to the successful accomplishment of becoming civilized. As a discipline with deep classical and humanistic roots, astronomy was considered an especially suitable subject for the education of gentlemen and ladies to cultivate literate and polite conversations.

The painting *A Philosopher Giving That Lecture on the Orrery* by Joseph Wright in 1766 is an iconic image of an astronomical demonstration in the context of polite science culture (fig. 1.1). A native of the English town Derby in the Midlands, Wright was acquainted with some leading scientific minds of the Lunar Society and was acutely aware of the changes brought by early industrialization. These interests caused him to choose subjects for his artworks portraying modern themes like science and technology.[25] The painting depicts the dramatic, romantic scene of an eighteenth-century lecture on astronomy. In the center of the painting, a "grand orrery" with an armillary sphere is exhibited. A small audience including children and adults surround the grand orrery, expressing emotions of awe and curiosity or making pondering gestures. The philosophical lecturer points out the models of heavenly bodies in a steady and authoritative posture. A man to the side takes notes on the lecture attentively. Because the number of attendees is so small, this scene more likely represents a private lecture at a country house. Nevertheless, it may not be an actual scene, although the artist might have based his depiction on real people. Researchers

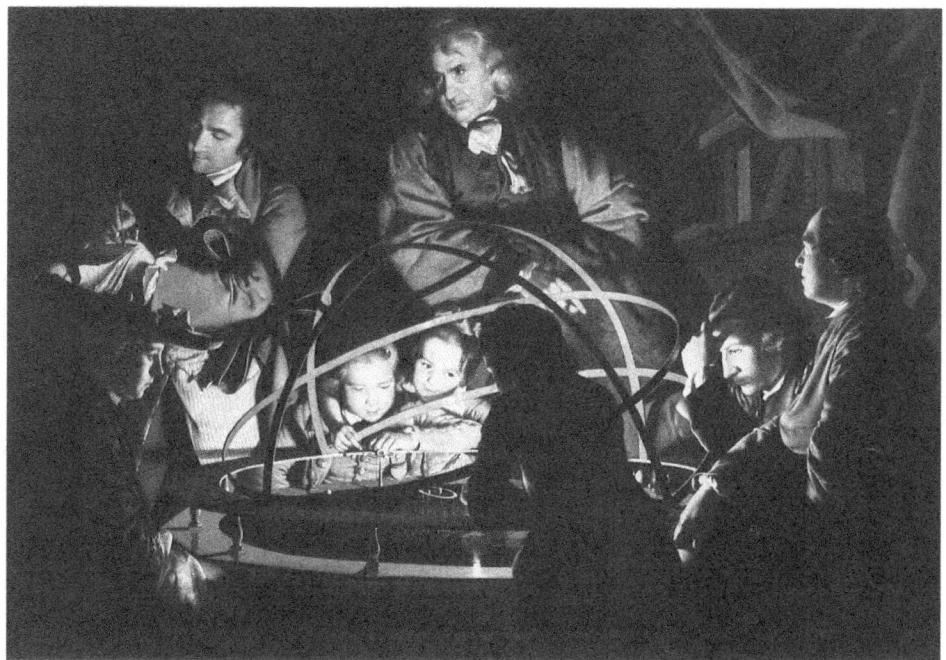

FIGURE 1.1. *A Philosopher Giving That Lecture on the Orrery, in Which a Lamp Is in Place of the Sun.* Joseph Wright, 1766. © Derby Museums and Art Gallery.

are still divided on the identity of the philosopher in the painting (or the model he is based on). For instance, Jonathan Powers claims that the philosopher was John Arden, while some claim the figure resembles James Ferguson or even a portrait of Isaac Newton.[26]

Another good example of polite science culture is Benjamin Martin's popular treatise *The Young Gentleman and Lady's Philosophy* (1759). This textbook was prepared for lay readers who had no prior knowledge of natural philosophy. Martin adopted the literary technique of dialogue to compose the narrative. In the book, a scientific lesson unfolds in a polite conversation between two fictional characters: Cleonicus, an educated young man, and his younger sister Euphrosyne. Cleonicus teaches knowledge of natural philosophy in various subjects to Euphrosyne (fig. 1.2). Although this is just a private one-on-one tutorial in a domestic scene, like contemporary lecturers, Cleonicus demonstrates experiments with many exquisite apparatuses and entertaining showmanship to capture the audience's attention. When teaching astronomy and its relevant instruments, the conversation went:

Cleon. Thus have we passed through so much of the Doctrine of the Sphere, and is necessary to enable you to have a right Understanding of the *Globes* and *Orrery*, which will exemplify and illustrate all those Matters, and make them familiar and easy to your Apprehension.

Euphros. Dear *Cleonicus*, you highly oblige and delight me; I think long e'er those Instruments come in Play: Pray, which are we to have next, the *Globes*, or *Orrery*?

Cleon. The Orrery, my *Euphrosyne*; I have provided one for that Purpose, which will afford you a pleasurable Entertainment for the next Leisure-Hours.[27]

It is obvious that Martin has attempted to create a polite conversation in this textbook. By using words such as "pleasurable entertainment" and "leisure hours," Martin attempts to tailor natural philosophy in the fashion of polite science and expects readers to connect philosophical instruments—in this case, orreries and globes—with desirable rational amusement. *The Young Gentleman and Lady's Philosophy* showcases the ideal enjoyment that Enlightenment science authors expected readers to gain from rational recreation.

Like many eighteenth-century science entrepreneurs, Benjamin Martin was also a polymath who engaged in multiple businesses across lecturing, writing, publishing, and the instrument trade. He was a versatile author of many astronomy and natural philosophy textbooks, published an English dictionary and a magazine of arts and sciences, and dealt in trading as an optical, mathematical, and philosophical instrument maker in Fleet Street in London after 1756. His most significant legacy to popular astronomy was a redesign of the orrery. The bulky grand orrery seen in Joseph Wright's painting was elaborate and expensive; it was made using state-of-the-art technique but it was impossible for every household to afford. Many philosophical lecturers, including Desaguliers, Ferguson, and Martin, had attempted to improve orreries. It was Martin's design of a portable orrery with a simplified "double-cone" wheelwork mechanism, which he described in detail in *The Young Gentleman and Lady's Philosophy*, that successfully transformed the instrument from a luxury curiosity to a more affordable commodity. Martin also designed different devices for specific functions: the "lunarium" was for showing the Earth-Moon sys-

FIGURE 1.2. The frontispiece to *The Young Gentleman and Lady's Philosophy*, Benjamin Martin, 1759. Image digitized by the Internet Archive, with funding from the Wellcome Library.

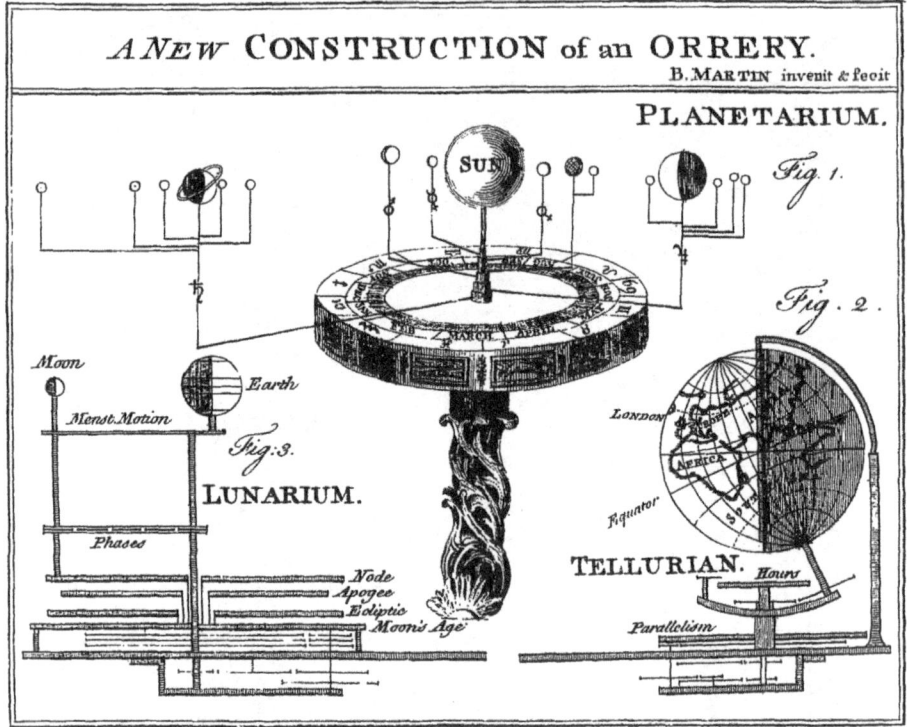

FIGURE 1.3. "A New Construction of an Orrery," a plate from *The Description and Use of an Orrery of a New Construction*, Benjamin Martin, 1771. Image digitized by Gale.

tem; the "tellurian" demonstrated the tilt and rotation of the Earth; and the "planetarium" was used to exhibit the whole solar system (fig. 1.3).[28] Each device could be mounted on or unmounted from the platform of the orrery as an attachment. Martin's portable design was later widely adopted and further improved by other artisans, such as the London-based instrument-making partners W. & S. Jones.[29]

Martin's contemporary, James Ferguson, was another important lecturer on astronomy in the mid-eighteenth century. A son of a farmworker from Banffshire in northeast Scotland, Ferguson was raised in a relatively humble background. Before his early thirties, he was employed as a handyman in many jobs, from house servant to limner of portraits. He was nevertheless a diligent autodidact and had kept up an interest in mechanics and astronomy. In 1743 he moved to London intending to start a career as a philosophical lecturer; there he invented astronomical devices for demonstrating celestial mechanics and exhibited them to the Royal Society. Ferguson also lectured widely throughout the country and his scientific reputation gradually grew.[30] When Samuel Johnson and his biographer James Boswell met Ferguson in 1769, Boswell designated Ferguson as "The self-taught Philosopher." Ferguson did not become a well-established commercial instrument maker like Martin; his interest and energy were more devoted to designing instruments rather than maintaining a shop. His scientific career was, however, much more successful than Martin's, as he had received patronage since 1761 from King George III with an annual pension of £50 and had gained institutional recognition. Ferguson was elected a fellow of the Royal Society in 1764 and a member of the American Philosophical Society in 1770. Martin, in contrast, never received the coveted title of a fellow and tragically went bankrupt in his late years after having poorly managed his business.[31] The difference in the outcomes of the two most eminent lecturers at the time underscores the commercial risks of natural philosophy lecturing ventures.

Both Martin and Ferguson were accomplished popularizers of astronomy and had published numerous plain and accessible works for lay readers. In particular, Ferguson's *Astronomy Explained upon Sir Isaac Newton's Principles* (1756) was an extremely influential and enduring bestseller that had been reprinted many times until the mid-nineteenth century. Even William Herschel, the astronomical hero later famed for the discovery of Uranus, was inspired by Ferguson's works in his early years as a professional musician and an amateur enthusiast of astronomy. Herschel's diary shows that he purchased a copy of *Astronomy Explained* in 1773, which was also the year he seriously turned to the pursuit of a career in astronomy.[32] He might also have attended Ferguson's lectures at Bath or Bristol, although there is no direct evidence of this.[33]

THE WALKERS AND THE DAWN OF STAGE ASTRONOMY

By the 1760s, Martin and Ferguson had made themselves accomplished lecturers of experimental philosophy and were both renowned for their prolific writings and instruments of astronomy. There were still more and younger rising stars on the horizon of the public lecturing trade. The key figures initiating the drastic transformation of popular astronomy toward the last two decades of the eighteenth century were Adam Walker (1730/31–1821) and his eldest son William Walker (bap. 1766, d. 1816). They put astronomy lecturing on the stage and turned it into a more dramatic celebration of knowledge with sensational visual and acoustic effects.

The patriarch and founder of this long-lived family business, Adam Walker, was born in Patterdale, Westmorland, a small village in the Lake District of northwest England.[34] Little is known about his parents and childhood, but it appears that he came from a humble background and had to make a living from scratch. Since his teens, he had been employed in various jobs as a worker, school assistant, and accountant. He received very limited education from school but developed considerable mechanical skills and learned mathematics from the jobs he undertook in his early years. He ran a school in Manchester in his early thirties, and then eventually decided to start an itinerant lecturing career. In 1766 he bought philosophical apparatuses from William Griffis.[35] He toured across northern England and southern Scotland. He also made friends of James Ferguson and Joseph Priestley, and was among the earliest lecturers to incorporate Priestley's discoveries of chemical properties of air into experimental philosophy lectures.[36] An intimate of several members of the Lunar Society, he was on good terms with its key figure, the inventor and businessman Matthew Boulton, who invited him to dine at Soho House when he lectured in Birmingham in the summer of 1781.[37] By that time, Walker had established a good reputation for polymathy and mechanical ingenuity. He then settled in London to fill a vacancy as an astronomical lecturer after the death of Ferguson.

William Walker was baptized at Kendal, Westmorland, in the same year that his father turned to itinerant philosophical lecturing. Adam Walker's experience of hard work and striving for self-education in the early years must have led him to consider how to improve his children's prospects. He groomed his eldest son to take over the lecturing business, ensuring that William traveled in higher social circles than he had been able to. William was educated at Eton College from 1778 to 1780, and had been taught Latin, Greek, and mathematics by private tutors before attending Eton. William Walker did not disappoint his father. The prodigy soon showed his talents, making his astronomical lecture debut with his father at Newbury in Berkshire in 1782 when he was only sixteen.[38] Adam Walker expected his son to be more than an ordinary commercial lecturer of astronomy. For this reason, he even accompanied William on the Grand Tour to Continental Europe between 1785 and 1787, with a view to observing the state of science and culture and forming "many respectable friendships."[39] In the next three decades, William Walker's lecturing business thrived until his premature death.

The invention that made Adam and William Walker famous, as well as the accoutrements that they applied to revolutionize popular astronomy, were the eidouranion, or the transparent orrery. Unlike its smaller portable kin such as what Benjamin Martin produced, the transparent orrery was intentionally designed to be displayed for a large theatrical audience. Nor was it like a conventional orrery, which was usually placed on a table, with the balls representing the planets orbiting horizontally to the ground. The disk of its rotating planets stood vertically, resembling a clockface perpendicular to the ground. As early as 1771, Adam Walker had mentioned his newly designed transparent orrery, describing it as "more like nature than anything of the kind (invented by the author)," in that year's syllabus.[40] A magazine article published many years later even claimed that Adam Walker had completed a model around the early 1760s.[41] Later, he named the device an "eidouranion." This curious name was derived from the Greek transliterations *eidó* and *ouranos*, literally meaning "image of the heavens" or "behold the sky." The eidouranion also seems to have been displayed in 1781 at the New Theatre in Birmingham, in 1782 at the Theatre Royal Haymarket in London, and again in October of the same year at Newbury, when the young William joined the lecturing business and made his debut.[42]

It is not odd that the exact date of the invention of the eidouranion was uncertain, for Adam Walker likely modified and improved its prototype through the years. It is also possible that the technology, not only the hardware but also the performance, had finally matured around 1782, as the eidouranion was frequently exhibited within a year of when young William Walker joined the troupe. In that same year of 1782, the Walkers also published the first edition of a little tract dedicated to the eidouranion, initially titled *An Account of the Eidouranion, or the Transparent Orrery*, which contained a general description of the machine and an outline of each scene in the show. The introduction began:

FIGURE 1.4. Portrait of William Walker. Stipple engraving by Ridley after a miniature by Miss E. Barratt. Published by Verner and Hood, London, 1798. © The Board of Trustees of the Science Museum, London.

This elaborate Machine is 20 feet [in] diameter; it stands Vertical before the Spectators; and its globe is so large, that they are distinctly seen in the most distant part of a theatre. Every planet and satellite seems suspended in space, without any support; performing it's [sic] annual and diurnal revolutions without any apparent cause. It is certainly the nearest approach to the magnificent simplicity of nature, and to its just proportions, as to magnitude and motion, of any Orrery yet made: and besides being a most brilliant and beautiful spectacle, conveys to the mind the most sublime instruction.[43]

This description was the same as the announcement in the London newspaper *Morning Herald and Daily Advertiser* on January 25, 1782. It became a typical account of the device for the advertisement.[44] The tract also served as a syllabus or a synopsis of the eidouranion lecture,

FIGURE 1.5. Group portrait of Adam Walker and his family, by George Romney, 1796–1801. Front row (left to right): Eleanor Walker (wife); Eliza Walker (daughter); and Adam Walker. Back row (left to right): William Walker; Adam John Walker (second son); and Deane Franklin Walker. Inventory no. NPG 1106. © National Portrait Gallery, London.

and would be reprinted for many editions in the next four decades. It seems that Adam Walker had handed over the management of the eidouranion lecture to his son by the early 1790s, as William Walker was specifically credited as the lecturer in the tract.[45] The portrait of William Walker published in the late 1790s, as figure 1.4 shows, also identified him as the "Lecturer on the Eidouranion."

Adam Walker had three sons (fig. 1.5). William Walker was the heir and partner of his father's scientific lecturing business, ran the eidouranion lecture, and took care of subsequent publications until 1816, when he died after a prolonged illness. The second son, Adam John Walker, graduated from Cambridge and became a clergyman, serving as rector and vicar of parishes on the Welsh border. The youngest son, Deane Franklin Walker (1778–1865), took over the family lecturing business after William's death, and continued the legendary lecture of the eidouranion.

The success of Walker's eidouranion attracted many imitators and competitors. Adam and William Walker had created not only a machine but also an unprecedented genre of a show on the stage. With this new "stage astronomy," astronomical lectures became independent from conventional experimental philosophy lectures, which now transformed from being one subject in the curriculum of natural philosophy to a distinct fixture that enjoyed popularity in the coming decades. We will further discuss the eidouranion and this theatrical turn of popular astronomy throughout this book.

Toward the turn of the nineteenth century, with the cultural and scientific foundation laid by public lecturing on natural philosophy, and the advent of various visual technologies such as the orrery and the eidouranion, popular astronomy lectures had become an especially attractive performance in the vein of Enlightenment rational recreation. Astronomy had long been deemed the sublime science, and there was no shortage of authors and lecturers to demonstrate the secrets of the universe. However, it was not until the theatrical turn of popular astronomy that the sublimity of the heavens and the rudiments of astronomical knowledge were able to be effectively conveyed to the wider audience. Ferguson, Martin, Walker, and many other philosophical lecturers in the eighteenth century contributed greatly to shaping this scientific trade and solidified the representation of astronomy as the sublime science. Their influence on subsequent popularizers of astronomy in the coming century was apparent and enduring.

In chapter 2, we explore the diverse scientific and entertainment venues in the country and its cities in the nineteenth century, examining the locations where astronomical lectures took place.

CHAPTER 2

GEOGRAPHY

Located at the eastern edge of Regent's Park in London, the Colosseum, an amusement facility famed for its enormously panoramic view of London and fine art exhibitions, enjoyed its glory years before the 1850s. The attraction, however, declined and eventually closed in 1864. In his analysis of the demise of the Colosseum, Richard Altick attributes the loss of appeal of the Colosseum to its location. He suggests that Regent's Park was relatively remote from London's center of amusements. The audience had a larger choice of shows in other more conveniently accessible places such as Leicester Square and the Strand.[1] The Adelaide Gallery in an arcade near Trafalgar Square had a more advantageous location. The bustle of its surrounding streets helped attract a great many curious crowds.[2]

Location was crucial, but so was architecture. James Secord compares the architectural characters of two institutions to highlight the political, religious, and cultural tensions in early Victorian Liverpool. The Liverpool Mechanics' Institution, where lectures on "useful knowledge" for the working classes were conducted through a Unitarian initiative, adopted a classical façade representing Enlightenment rationalism. In comparison with the dissenters' liberal vision, the Anglican Church preferred a more traditional value. The building of the Collegiate Institution, which was sponsored by the Church of England to provide "suitable" education in science and religion to sons of middle-class elites, was designed in Oxbridge's favorite Tudor Gothic style. The significant contrast between the two institutions was, in Secord's words, "built into stone."[3] Architecture was a symbolic rendering of space for an implicit expression of values and ideologies.[4]

Spatial features such as location and architecture all speak to the fact that place matters. Nineteenth-century astronomy lectures took place in a variety of venues, the spatial characteristics of which shaped lectures into distinctive forms. London's West End theaters, where many Lenten astronomy lectures were performed, greatly differed in function, location, and architectural space from scientific establishments such as the Royal Institution. Sites are not the only geographical differences among myriads of lectures. Regions, whether referring to a fixed area on the map or a relative division between the center and the periphery, are also a significant geographical factor in shaping a lecture's locality. The provincial towns in Northern England or Scotland, where Deane Franklin Walker frequently toured, were incomparable to the splendor of London's Theatreland. The character of these diverse places invites serious thinking about the distinctive effects of space on the practice of astronomical lecturing.

This chapter explores the geographical features of astronomical lecturing: where astronomy lectures took place and, moreover, how they were presented. The history of science over the past two decades has taken a spatial turn. Historians have attempted to think "geographically" about the production and movement of scientific knowledge. Scholars have considered various notions of geographical scales such as cities, sites, regions, and territories.[5] Rather than a single, unified rationality in its practice, science is performed in diverse localities in which spatial conditions can create very different outcomes.[6] As David Livingstone and Charles Withers rightly suggest, geographical approaches could prompt researchers to ponder two sets of key questions: the making and meaning of science *in place*, and the movement *over place*.[7] Places where scientific knowledge is generated and communicated are not merely stages on which actors perform. They shape actors' relationships, facilitate and constrain actors' practices, and affect the theories conceived by actors. In addition to context-situated science, the movement of scientific knowledge through different physical or mental spaces is noteworthy. How does science, as the diffusionist model perceives, spread from "expert" to "lay" audiences? How does science travel within and between communities? How does it move from place to place? These questions lead us to think about the changing state of science in motion, or as Secord calls it, knowledge in transit.[8]

The spatial turn in the history of science has motivated researchers to study places outside laboratories and field research stations, the two privileged "default" sites for the making of scientific knowledge from the twentieth century to the present day. Labs and field are, to use Thomas Gieryn's term, "truth-spots" that conventionally lend credibility to scientific claims.[9] Science also happens in many other settings beyond those privileged truth-spots. These "unconventional" sites, including expositions, zoos, aquariums, botanical gardens, royal courts, parlors, and even cafés and pubs, are also locations where scientific knowledge is produced and propagated. The examination of spaces also prompts historians to rethink some familiar sites of science including normal truth-spots. For instance, although laboratories are synonymous with scientific research nowadays, they did not become essential and widespread until the mid-nineteenth century. Museums, in contrast, were important sites associated with the production and preservation of knowledge before and throughout the nineteenth century.[10] Exploring spatial characteristics and their historical, social, and cultural contexts was therefore helpful to better understanding the changing roles of specific places or sites over time.

Astronomy lectures in the nineteenth century were held in a wide diversity of sites too. This chapter's examination of these various cases argues that the physical space of a site was a significant factor in shaping the format and style of and the audience for astronomy lectures. Furthermore, studies of astronomy lecturing should emphasize both visual and acoustic cultures that were shared by contemporary performers as well as spectators—in other words, a science feast for the eyes and ears. Scholars using geographical approaches to examine different places of scientific practices and ideas in recent years have often focused on the spatial effects of techniques for visual display and entertainment. Some researchers remind us about the importance of sound and verbal renditions in the execution of scientific speeches. As Diarmid Finnegan puts, science had to "sound right as well as look right to retain its place as part of intellectual culture" in Victorian Britain.[11]

Theaters as locations were often overlooked by historians, yet they best exemplify spatial effects on both visual and acoustic experiences for astronomy lectures. The Walkers made a crucial contribution to the early development of theatricality in astronomy lecturing. Astronomy lectures performed in theaters had many features distinct from those delivered at other venues. The benefits and constraints of theaters offered researchers vivid cases for comparison with other spaces.

The following exploration of the landscapes of spaces for astronomy lecturing begins with an overview of the cultural and leisure milieu of London, the British capital and the most affluent metropolis. Cities like London are variable arenas for addressing notions of geographies of science; urban resources can result in certain effects that are hard to reproduce or copy in other places.[12] We will visit different metropolitan sites where the stories of the heavens were told, and we will follow the footsteps of several itinerant lecturers to other places in the British Isles, reaching even very remote places in the country.

The Metropolis as a Cornucopia of Amusements

"When a man is tired of London, he is tired of life; for there is in London all that life can afford." So said Samuel Johnson to his biographer and friend Boswell in 1777, this quotation timelessly expresses the affluence and self-contained pride of the metropolis. As the capital of a wealthy and expanding empire in the early nineteenth century, London was full of attractions. It is not difficult to imagine the dazzling display of hustle and bustle observed by any visitor first arriving in London. Krystyn Lach-Szyrma, a Polish intellectual who first visited London in 1820, left his thoughts in a memoir, where he wrote, "Having known that I was approaching the biggest and the wealthiest capital city in the world, I was overcome by unusual emotion."[13] Lach-Szyrma strolled the streets of London in utter bewilderment, later recalling that he "did not know where to go" and "did not care." He described his first impression of London: "I was walking in the streets and squares, over the bridges, looking at shops, vehicles, tombs, dresses, monuments and people. Whenever I wanted to stop somewhere for longer, crowds of people, going in the same direction as I was, pushed me and took me with them as though in a flood and as they wanted to tell me that what I had seen was nothing compared to what I was to see further on."[14] The metropolis indeed had its own glamour to impress a visitor.

The abundant resources that London could offer a visitor consisted of not only material fortune but also cultural and scientific activities.[15] From the Egyptian Hall in Piccadilly (1812–1905) to the rotunda Panorama in Leicester Square (1793–1863), the metropolis had so many shows and exhibitions to indulge in. The Royal Institution in Albemarle Street and many other learned societies also provided "rational recreations" for science lovers. Such a cornucopia of amusements made London a vibrant marketplace. Scientific lecturing ventures were a part of this marketplace; the entrepreneurs in this trade competed with each other and with other amusement businesses.

Readers can imagine the urban life within which astronomy lecturing was embedded by joining a virtual trip to early nineteenth-century London. On this journey we will "visit" various London tourist attractions of the time. A trip on paper is an effective and fun way of setting the scene. Scholars have indulged in this literary technique for reenacting the experience of visiting metropolitan places. Richard Bellon, for example, shows a suggested route to the Crystal Palace at Hyde Park according to a contemporary guidebook. Bernard Lightman, too, walks through various scientific sites in London as a traveler.[16] Following their steps, at this juncture, we will embark on a guided exploration of the city.

Two kinds of primary sources can help historians to re-create the vibrancy of the city: visitors' reminiscences of their London experience and contemporary guidebooks to the metropolis. Lach-Szyrma's memoir, for instance, is a remarkable example of the former that allows us to see through the eyes of a foreign stranger who shares fresh personal insights. Lach-Szyrma was more than just an intelligent tourist. As a member of the Polish educated elite who had a radical vision of political reforms for his home country, Lach-Szyrma's memoir covers many distinct aspects of the metropolis, ranging from the legal system to popular amusements. Despite his being dazzled by London's excitements, he made important observations about British institutions and society. The other primary source, guidebooks to London, provides another perspective, the one that locals wanted to present to visitors. Guidebooks flooded into the nineteenth-century publishing market in various forms, from pocket-size handbooks to thick gazetteers. These publications often listed a variety of landmarks, sites, and activities, with informative descriptions, and sometimes including the author's personal recollections. They offer a window on the fashionable attractions and entertainments for contemporaries. *London Lions for Country Cousins*

GEOGRAPHY

▶ FIGURE 2.1. A map of London in the 1820s showing the locations of theaters and scientific institutions. The image is reproduced from *A New Map of London and Its Environs*, by Mr. Thompson (cartographer), Reeves and Hoare (publisher), 1822. By courtesy of Harvard Map Collection, Harvard College Library, https://hgl.harvard.edu/catalog/harvard-g5754-l7-1822-t5. Numeral and alphabet symbols added by Hsiang-Fu Huang.

Scientific institutions

A. Royal Institution
B. London Institution
C. Russell Institution
D. London Mechanics' Institution
E. Surrey Institution

Theaters

1. King's (Her Majesty's, after 1837)
2. Haymarket
3. Adelphi
4. English Opera House (Lyceum)
5. Strand
6. Princesses's
7. Drury Lane
8. Covent Garden

and Friends About Town (1826) by Horace Wellbeloved, for example, was representative of the guidebook genre in Lach-Szyrma's time.

London Lions is in many ways a well-crafted guidebook. The book's title is intended to be humorous, as the author boasts in the preface: "Our COUNTRY COUSINS, when they come to see the '*London* LIONS,' will have only to put this Volume in their pockets, and by its direction, will be led from place to place, requiring no other guide."[17] The playful title alludes to the famous tale "the town mouse and the country mouse" from Aesop's fables. Moreover, the term "lions" had a particular connotation in nineteenth-century social culture: "lion" could describe an accomplished individual, often an influential literary author or a social celebrity. The opposite concept of a "lion" connoted an unfavorable person—a "bore," often depicted punningly as a boar. A lion could be the focus of attention at a party or in conversation; on the contrary, a bore's unpleasant talk and manner could drive people away. This animal metaphor represents an important social protocol about what was right or wrong behavior in the oral culture of nineteenth-century society.[18] The book's title and preface therefore conveyed an obvious message to readers that they could rely on *London Lions* not only for travel information but also for elegant conversation pieces, as the book contained the most fashionable subjects of metropolitan attractions and novelties. *London Lions* collected information about prominent buildings, locations, and organizations in the metropolis and especial-

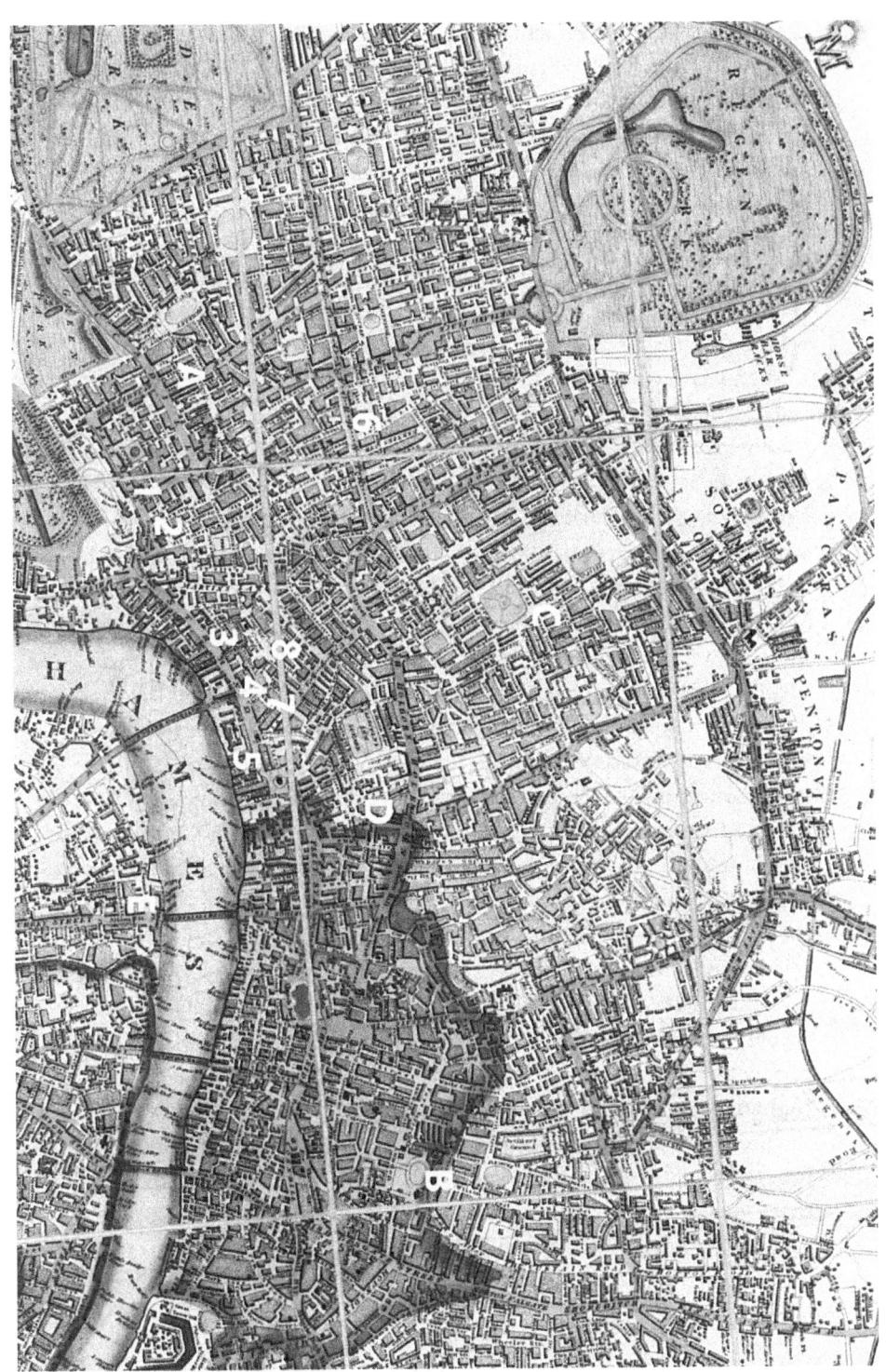

ly emphasized improvements and amusements. Each entry elaborated the place's historical and cultural background. Besides being practical, *London Lions* was also artistic and collectable. It contained more than twenty exquisite engravings of landmark buildings, events, and urban landscape. The relatively inexpensive price of the book, 5s. 6d., was also appealing. Overall, *London Lions* was an affordable, quality guidebook for middle-class readers.

So, what amusement attractions did *London Lions* recommend? Before we start our sightseeing, a map showing the city's streets is essential. Figure 2.1 shows a map of central London and the locations of the places we are about to introduce. We start at Regent's Park, at the northern edge of the urban area in the mid-1820s. Unsurprisingly, Wellbeloved praised the pastoral beauty of Regent's Park, and another attraction next door, the Diorama, drew visitors' attention with its artificial beauty. The Diorama at Regent's Park offered spectators a sensational visual experience. It was a scenic display of flat paintings with the illusion of depth—a nineteenth-century version of a 3D film.[19] With the sophisticated application of adjustable lighting, the Diorama could create dramatic effects that immersed the audience in the scenes. The Diorama's popularity was evident, as Wellbeloved devoted six pages to describing the various scenes on display at the Diorama, with three illustrations showing the most popular ones.[20] The *Mirror of Literature*, one of the earliest cheap weeklies in England, also published a frontpage feature article on the scene of Rosslyn Chapel when it debuted at the Diorama. It praised the scene as surpassing others previously presented at the Diorama, and stated it would be "one of the most attractive features of the most fascinating exhibition ever opened in London within our knowledge."[21] Lach-Szyrma visited the Diorama in 1823 and he was overwhelmed by the uncannily realistic scenes of the interior of Canterbury Cathedral and the Swiss valley of Sarnen: "As I was looking at the two views it seemed to me that, as though by means of supernatural power, I had been transported into the precincts of the cathedral I had already visited and into the most beautiful valley in the world in order to feast my eyes once again upon these views."[22] *London Lions*, however, does not mention the Colosseum, the Diorama's future neighbor and rival. The Colosseum was then under construction and did not open until 1829, three years after the publication of *London Lions*.

As we leave Regent's Park to follow Portland Place and Regent Street on the south side of the park, heading south toward the Thames, we soon arrive at Mayfair and Soho. These two areas are the heart of the West End of London, where many tourist attractions were to be found in the Leicester Square and Piccadilly vicinity. By 1826, the Royal Institution had already been in Mayfair for decades, but *London Lions* neglects this scientific landmark because it was for the elite, and only wealthy members could get in. Perhaps the Egyptian Hall in Piccadilly was more fun and more accessible to tourists. The Egyptian Hall, known for the building's Egyptian Revival–style façade and interior and sometimes referred to as Bullock's Museum, was erected by the naturalist and antiquarian William Bullock in 1812 to house his collections. It was Londoners' longtime favorite place to see exhibitions of curiosities. When *London Lions* was published, the exhibition at the Egyptian Hall that attracted the most public attention consisted of artifacts and specimens from Mexico, which Bullock had collected during his expedition in 1823. Visitors could stroll around altars and colossal idols of the Aztecs in the exhibition.[23] This was not the only exhibition to capture the attention of visitors; over its nearly century-long history, the Egyptian Hall had hosted a variety of attractions, including zoological rarities from the Cape of Good Hope, Mr. Faber's "speaking-machines" automatons, and the artifacts of "Iowy" (Iowan) Native American tribes; the hall resembled a strange hybrid of natural history and technology museums.[24]

Many minor exhibitions were presented around this area, besides those at the Egyptian Hall. For example, the Linwood Gallery in Leicester Square was famous for its exhibition of tapestry; the Royal Armoury and Automatons in Haymarket, where the legendary French emperor Napoleon's firearms were on display, along with the performance of musical automatons. For those fearless and ruthless spectators who "require the most repulsive anomalies of nature to excite their palled sensations," *London Lions* considers the unsparing exhibition of the "Living Skeleton" at Pall Mall as a perfect fit.[25] The Living Skeleton was an emaciated Frenchman, Claude-Ambroise Seurat, who toured across Europe as a freak show attraction and visited London in 1825. His extraordinarily low body weight also attracted the attention of artists and medical establishments.

London Lions did not include another important exhibition site in the vicinity of the West End: The National Gallery of Practical Science, popularly known as the Adelaide Gallery, was not mentioned, as the gallery was just opened in 1832, six years after the publication of *London Lions*. Located on the north side of Lowther Arcade in Adelaide Street, between Charing Cross and Covent Garden, the Adelaide Gallery would become a strong rival of nearby exhibition places including the Egyptian Hall. It was established by the American inventor Jacob Perkins, who invented an automatic firearm "steam gun" and tried to sell this new weapon to the military but failed. Perkins nevertheless found an alternative way to earn cash from his inventions, by building an exhibition showing his ingenious mechanical inventions or those of others. As one of its advertisements described itself, the gallery's unique features were based on "novelty and amusement blended with instruction." In its early days, in addition to Perkins's steam gun, the gallery also exhibited an oxy-hydrogen microscope, several working models of steam engines and steamboats, and an "electrical eel in full life and vigour, the only specimen ever brought to England."[26] Before its final closure and transformation into a marionette theater in 1852, the Adelaide Gallery had developed into an amusement hall for scientific and variety shows, which combined "the astonishing powers of electricity with the wonders of optical illustration, scientific lectures daily delivered, and introduction of promenade concerts."[27] The gallery's initial ambition to promote the arts and manufactures by their connections with science was soon replaced by another younger rising star, the Royal Polytechnic Institution, founded in 1838.

Astronomy had its own prestigious place among many entertainments in the marketplace. Mr. Walker's eidouranion (at the King's Theatre when *London Lions* was published) and Mr. Bartley's "uranologia" at the English Opera House occupied the place of the very first entry in *London Lions*. Wellbeloved praised: "Of those optical exhibitions of the higher class, which have had their combinations made by men of science, and which have been perfected by the aid of painting and mechanics, the eidouranion of Walker, and the improved lecture of Bartley, well deserve to rank the foremost."[28] D. F. Walker and George Bartley (ca. 1782–1858) were the two brightest stars in the shows of stage astronomy when the *London Lions* published, but soon Charles Henry Adams (1803–71, commonly known as C. H. Adams) and other newcomers would also compete together in the Lenten arena. Apart

FIGURE 2.2. "Pit, Boxes and Gallery," by George Cruickshank (draftsman), June 25, 1836. V&A: S.382-2009. © Victoria and Albert Museum, London.

from the eidouranion and the uranologia, *London Lions* introduces another astronomical exhibition, Busby's Orrery, a self-moving hydraulic machine invented by the English architect Charles A. Busby. It had the same effect as other self-acting orreries but was driven by hydraulics rather than clockwork. The machine was on display in a London exhibition room (though *London Lions* does not mention the exhibition's location), with a circular reservoir placed in the center of the room for demonstrating the hydraulic orrery. Wellbeloved remarked on the intriguing device's ornamental value and recommended Busby's Orrery to his readers for its promotion of "a love of science and general astronomical information."[29]

London Lions did not cover theaters, yet theaters occupied a key place in the metropolitan amusement market and cultural life. According to *Curiosities of London* (1867) by John Timbs, another more comprehensive guidebook published decades after *London Lions*, London had twenty-three "major" theaters.[30] These major theaters had been licensed by the Lord Chamberlain to present any kind of drama. Most of them were in the West End, including the famous Drury Lane, Covent Garden, and Haymarket theaters. On both sides of Trafalgar Square, along the broad Strand or Haymarket streets, many major theaters were located. London's theatrical performance venues were nevertheless not confined to the West End, as many minor theaters were scattered across the East End and the fringe of the metropolis. A parliamentary committee in 1866 reviewed theater licensing, and its report showed that theaters outside the West End made up 63.7 percent of the total capacity of London theaters, and many additional kinds of performance entertainments at music halls and penny gaffs appealed to the lower classes.[31]

The spectrum of nineteenth-century theaters was as wide as the diversity of society. Although a theater might attract a particular class of audience, it was possible to observe a miniature society on the same occasion. Most West End theaters divided the auditorium space into tiers of seating—such as boxes, pit, and gallery—like travel classes on public transport. A comical illustration "Pit, Boxes and Gallery" by the caricaturist George Cruickshank best depicts this social makeup in a theater (fig. 2.2).[32] Admission to a private box was the most expensive; those who could afford one were charged 1 guinea, which was about the weekly salary of a general shop clerk. A box seat was cheaper, but still usually cost more than two to three times the cheapest gallery ticket. Boxes were also a showcase for the rich to flaunt their fashion tastes, as Lach-Szyrma described: "[The boxes] were upholstered with a crimson cloth whose colour flattered the ladies who displayed themselves superbly, especially in the first row of boxes where only formally attired persons are admitted. Men attend plays clad in black tailcoats, stockings and boots with white scarves round their necks, for colorful and motley garments are deemed vulgar in London." On the contrary, the gallery on the top floor of the auditorium was often occupied by the poorest and rudest audience. Lach-Szyrma described the gallery audience as the "mob" that "hovers above the entire congregation." They would whistle, hiss, howl, and boo without hesitation, and occasionally "dirty jests, loud laughter and national songs can be heard."[33] James Grant also described the chaos of the lower-class audience crammed in the upper gallery in *The Great Metropolis* (1837), whom the contemporaries satirically called the "gods" of the gallery.[34] Although the chaotic situation Lach-Szyrma and Grant witnessed might not have happened every day, it reflects how diverse the background of audiences could be in a nineteenth-century theater.

Scientific institutions and learned societies were other kinds of sites that *London Lions* did not cover because of the guidebook's focus on amusements. The early 1820s, when Lach-Szyrma first visited London, was nevertheless a period during which London's scientific lecturing trade thrived and numerous scientific establishments expanded.[35] As a patriotic intellectual who was eager to push for political reforms in his home country, Lach-Szyrma certainly noticed the development of scientific institutions and their educational promise in Britain.[36] He mentioned several scientific institutions and learned societies in his reminiscences, including the Royal Society, the University of London, and the Royal Institution, which he described as "a kind of technical school." He was especially excited about those establishments that aimed to promote useful knowledge among craftspeople, such as the London Mechanics' Institution and the like. He praised such initiatives as "a huge step towards further progress in industry and crafts, as well as in morality" because it could direct artisans to refine crafts and promote inventions in their spare time, away from harmful entertainments such as consuming alcohol.[37] By the mid-Victorian era, scientific institutions and learned societies had proved to be an enormous force in the development of British science. Many of them were recognized by the public as a part of cultural life, and guidebooks could no longer neglect them. John Timbs, for example, wrote entries for the Royal Institution and the London Institution in *Curiosities of London*. He praised the London Institution's book collection in its library as "one of the most useful and accessible in Britain," and approved of the Royal Institution as "the workshop of the Royal Society" because of the brilliant discoveries made by Humphry Davy and Michael Faraday in its laboratory.[38]

The British capital was fascinating in the eyes of a foreign visitor and the authors of two guidebooks. Their accounts represented just a few examples in the vast literature yet were sufficient to inform modern readers about the richness of the metropolis as a cornucopia of entertainment and cultural life. Astronomy lecturing thrived and competed with other entertainments in this vibrant cultural environment. The various venues I describe here are not merely the setting of astronomy lecturing, for the physical spaces and the cultural context of these sites shaped quite diverse types of lectures. Next, we examine in detail each of the sites in which astronomy lectures took place, from theaters, scientific institutions, and mechanics' institutes to assembly places in rural areas.

Astronomy in Theaters

Long before he was a celebrity at the Royal Institution, Michael Faraday was aware of the importance of showmanship. When he began working there as an assistant in February 1813, he attended many scientific lectures, from which he learned lessons that would later benefit his lecturing. Faraday summarized a few observations in letters to his friend Benjamin Abbott; these observations were full of profound insights about not only the speakers' oratorical skills but also the space and technical aspects of the lectures. For example, he noticed the spatial features of the theater and compared these with other venues:

> The theatres in a large way have one advantage i.e. in the site of their stage lamps which illuminate in a grand manner all before them tho at the same time they fatigue the eyes of those who are situated low in the house but tho Walker has shewn in the most splendid and sublime manner that Astronomy may be illustrated in a way the most striking by artificial light yet from what little I know of these things I conceived that for by far the greater part of Philosophy day light is the most eligible and convenient.[39]

As Faraday remarked, astronomical lectures were distinct from others because of their need for a darker environment and thus artificial lighting. There was no better venue for an astronomical lecture than a theater. Faraday's favorite scientific lecture venues, as he indicated in the same letter, were the Royal Institution, the Automatical Theatre, and the Theatre Royal Haymarket. He commented that these were venues in which "I have seen company and which please me most."[40]

Faraday's account reminds us that theaters were a significant site of scientific lecturing in the early nineteenth century, especially astronomical lectures. Scholars have identified a variety of sites for the production and dissemination of scientific knowledge, yet theaters have received less attention. Public lectures inside theaters remain relatively obscure in the scholarship of the history of science, despite their extensive influence on contemporary audiences.

GEOGRAPHY

The "Walker" praised by Faraday for illustrating astronomy in the most splendid and sublime manner was William Walker. At the time when Faraday began his assistantship at the Royal Institution, William had already taken over his father Adam Walker's lecturing business. As a seasoned lecturer who carried on the family legacy in this trade, William paid attention to exactly *how* he presented the material. In addition to the theater itself, the organization of the lecture also drew Faraday's approval:

> The only instance in which I have seen a Lecturer succeed in occupying the attention of his audience for a time eminently longer than an hour was at Walker's orrery in which the subject has occupied time to the amount of two or three hours[.] But here we have peculiar attendant circumstances from the nature of the place itself (a theatre) we expect to remain there a considerable time & tho the subject differs from such as usually draw us there yet we in part associate the ideas together—Again Mr. Walker very judiciously leaves the audience at intervals to themselves during which time they are entertained by harmony well suited to accompany such a subject by these interruptions he allows the minds of his company to return to their wonted level.[41]

Faraday claimed that the spectators' perception of a theater and the well-designed musical intermissions were Walker's secrets for keeping the audience's attention. Thus, the venue and the format of the lecture were two crucial factors distinguishing Walker's talks from those of other lecturers.

Theaters possessed distinctive facilities that were difficult to find in other venues: stage lighting, orchestra pits, extensive platforms for accommodating large apparatuses, and capacity for hosting an exceptionally large audience. Large capacity and voluminous space were especially attractive features for lecturers in business. Take, for example, the Lyceum Theatre—also known as the English Opera House—where Walker, Adams, and Bartley had all performed. In figure 2.3 showing the ground plan of the Lyceum Theatre designed by the architect Samuel Beazley in 1816, we can clearly see the scale of the capacious stage and the slightly elongated horseshoe-shaped auditorium. Beazley built the Lyceum twice: first in 1816 and then in 1834 after a fire destroyed the theater. The size of the Lyceum Theatre was 51 feet "long" from the curtain to the front boxes and 35 feet "wide"

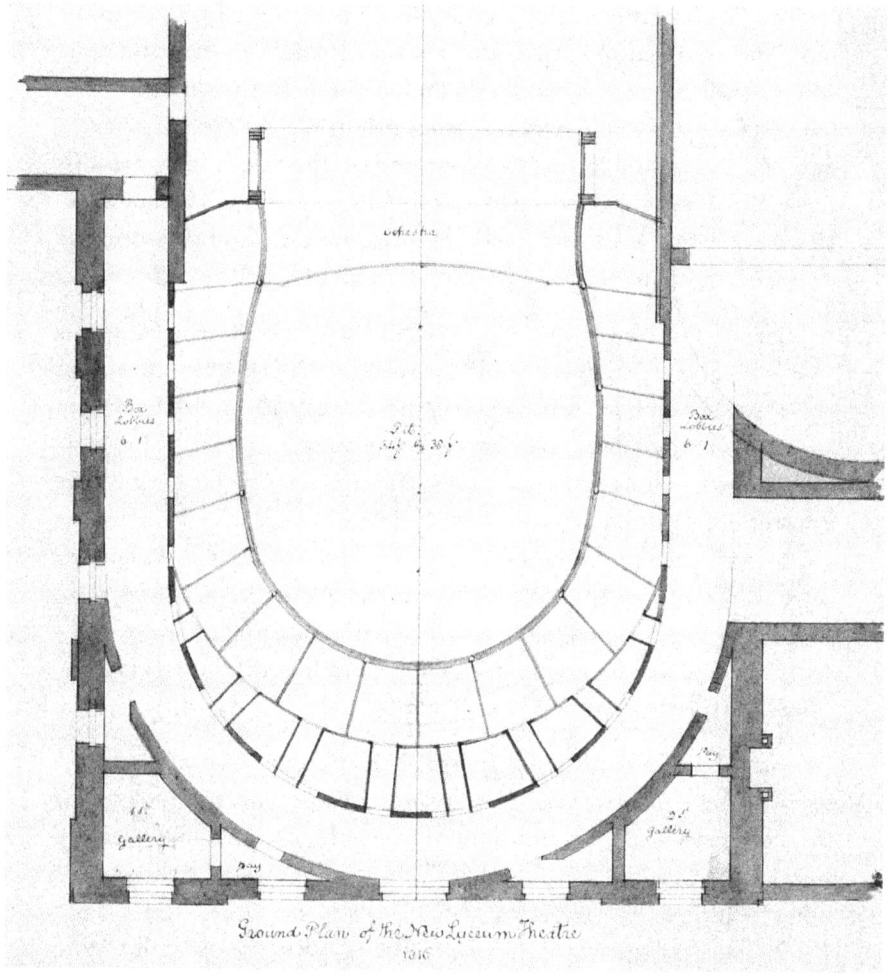

FIGURE 2.3. Ground plan of the new Lyceum Theatre, by Samuel Beazley (draftsman), 1816. The space at the top of the plan is the backstage and stage. V&A: S.396-1989. © Victoria and Albert Museum, London.

across the pit.[42] Although there are no accounts specifying the capacity of the Lyceum before the fire, the reconstructed theater could hold 1,490 spectators in total.[43] This number is probably not far from the old building's capacity. A West End theater could usually host over 1,000 people, sometimes reaching even 2,000 or more.[44] In contrast, the average attendance at a Royal Institution event—at its mid-nineteenth century peak—was between 400 and 500, less than half of a major West End theater's capacity.[45]

Size was certainly not everything. When it came to choosing a theater that best suited the performance, the bigger one was not always better. Beazley, who built and furnished many London theaters, including the Lyceum and Drury Lane, considered the best theater size for hearing and seeing a performance to be no more than 50 feet wide and 55 feet long from the curtain to the front boxes. These criteria were for the presentation of historical plays and tragedies, which require long processions.[46] Astronomy lectures did not require such a large space and, therefore, a smaller theater was sufficient to achieve satisfactory effects. Theaters that were frequently used by C. H. Adams, Bartley, and Walker, were usually medium-size ones (compared to the largest two theaters, Drury Lane and Covent Garden) with a capacity of around 1,500 people.

When considering the balance between space and audience size, the limitations of acoustic effects were sometimes even more important than that of visual effects. Visual effects could be achieved with large apparatuses or transparencies prepared in advance, but vigorous training was required to speak loudly and clearly in the vast space of a theater. Astronomical lecturers might not need to sing, but they certainly needed to have an eloquent voice in an age without microphones. In addition to beautiful scenery and simple elucidation, oratorical skill was another criterion reviewers used to judge a good onstage astronomy lecture. William Walker, for example, was praised by contemporaries for his clear and interesting "manner of elocution."[47] Faraday especially noted Walker's ability to occupy the attention of the audience. George Bartley, who dominated the Lenten fixture at the English Opera House during the 1820s, was also highly praised as "the best prose speaker" compared to other famous elocutionists, including the actor Stephen Kemble and the elocution tutor Benjamin Smart.[48] Bartley's success in Lenten astronomy lectures was also attributed posthumously to his "fine voice and perfect elocution."[49]

Music was the icing on the cake in addition to a good delivery. As Faraday observed, the arrangement of musical performances during intermissions was Walker's trick for relaxing the audience. Likewise, intermissions and musical performances were strategies to provide a break for the lecturer—after all, an astronomy lecture in a theater usually lasted for two to three hours. The music chosen for the performances on such occasions was also suitable for the themes of astronomy or Lent. For instance, the air *Holy, Holy, Lord* by George Frideric Handel and the duet *The Heavens Are Telling* from Joseph Haydn's oratorio *The Creation*, had been put into the Walkers' programs.[50] These arrangements of religious hymns as musical interludes helped render the subjects sublime and morally uplifting, as they praised the greatness of Creation and appealed to the audience's artistic and religious sensibilities.

In all respects, astronomy lectures in theaters were not just about scientific knowledge, but more like entertainment that blended instruction with moral education and religious sentiments. In terms of execution, they also resembled performances that combined various artistic media including visual display, music, and speech. The artistic element of this trade aimed to evoke sensational feelings in addition to merely conveying a knowledge of physical science. Such renditions were welcomed by some contemporary critics such as Horace Wellbeloved, who highly praised the potential of this distinctive type of astronomical show. In *London Lions*, he asserted that an elaborate arrangement of machinery, a correct presentation of the science, and a proper display of distances and magnitudes, were *not* the "ne plus ultra" of astronomy lecturing. Further cooperation between Urania and Terpsichore, the Muses of astronomy and choral song, he argued, was an ideal union for successful presentations of the science of astronomy, using the ornamental arts as auxiliaries.[51]

Theaters thus became a common venue for hosting popular astronomy lectures in early nineteenth-century Britain. While William Walker and his youngest brother D. F. Walker demonstrated the eidouranion at the Haymarket and the English Opera House before 1830 (fig. 2.4), their longtime competitor R. E. Lloyd also exhibited his dioastrodoxon in theaters across Great Britain. Lloyd's lecture circuit included places such as London's Haymarket Theatre, the Music Room in Oxford, and the Caledonian Theatre in Edinburgh.[52] Many imitators were entering this show business, and some of them adopted stage astronomy as a sideline. For example, James Howell, a clerk in the City of London during the daytime and a showman at the Strand by night, delivered astronomy lectures at the Adelphi and Her Majesty's Theatres in the 1830s.[53] C. H. Adams, the most celebrated Lenten astronomy lecturer in the mid-nineteenth century, performed over thirty consecutive years beginning in 1830. Adams's stage

FIGURE 2.4. "The Proscenium of the English Opera House in the Strand (Late Lyceum) as it appeared on the Evening of the 21st March 1817, with Walker's Exhibition of the Eidouranian," by E. F. Burney (artist), I. Stow (engraver), Robert Wilkinson (publisher), October 11, 1817. V&A: S.176-1997. © Victoria and Albert Museum, London.

sideline seemed more splendid than his everyday career as a school headmaster.

The most impressive case of astronomy lecturing in the theater as a "sideline" was, understandably, performed by a professional actor. The veteran comedian George Bartley (fig. 2.5) delivered Lenten lectures on astronomy at the English Opera House in the 1820s. Bartley regularly played at Covent Garden in winter and at the English Opera House in summer. He was especially famed for performing comic older men such as Falstaff and Sir Toby Belch in Shakespearean plays.

FIGURE 2.5. Portrait of George Bartley in his elderly age, lithograph by Richard James Lane, 1846. Inventory no. NPG D21693. © National Portrait Gallery, London.

FIGURE 2.6. Portrait of Samuel James Arnold, lithograph by William Pengree Sherlock, around 1830–40. Printed by Lefevre & Kohler. British Museum, inventory no. 1893,0123.2. © The Trustees of the British Museum. All rights reserved.

FIGURE 2.7. Playbill of George Bartley's lecture at the English Opera House during Lent 1827. RAS: Add MS 88: 7. Photographed by Hsiang-Fu Huang. By permission of the Royal Astronomical Society, London.

The idea for lectures on celestial topics given by a celebrated comedian was actually from the proprietor and manager of the English Opera House, Samuel James Arnold (1774–1852), who also penned the lecture's script (fig. 2.6). Bartley's lecture was therefore a piece of collaborative work between the two theater professionals. Although Bartley's astronomical sideline only took place during Lent and lasted for less than a decade—between 1821 and 1828—it was evidently successful (fig. 2.7).[54] Despite the lack of records for box office or attendance, contemporary accounts reveal critical acclaim. For example, the optician and popular author William Kitchener recommended that his readers pay a visit to Bartley's lecture, in which "the most beautiful and perfect Orrery ever exhibited" was shown, and stated that Bartley "well deserved the fame he has acquired, by the impressive manner in which he delivers his illustrations of these sublime subjects."[55] The popularity of Bartley's lecture was even referenced in satirical poetry. A drollery called *Love and Lunacy*, written by Thomas Hood in 1836, mentioned Bartley's lecture. In this drollery the protagonist Lorenzo, a fictional young man who loves science, is in agony because of his girlfriend's ignorance about astronomy. Lorenzo blames his girlfriend's cluelessness on their being in the wrong place on their date:

> Fool that I was, in my mistaken zeal!
> I should have led you—by your leave and pardon—
> To Bartley's Orrery, not Covent Garden![56]

This anecdote illustrates that Bartley's name was well connected to his Lenten sideline, even after the show had been closed for nearly a decade.

Lent was a particular season for astronomy lecturing in theaters. A contemporary journalist once described the phenomenon in a humorous way as "Astronomical Mania."[57] Several factors contributed to the prevalence of Lenten astronomical lectures. One apparent reason was the ban on dramatic performances during Lent. Because Lent was an important season in Christian tradition, many religious customs remained in place during Lent in early nineteenth-century Britain, including restrictions on theater operation. Theaters were closed on Wednesdays and Fridays throughout the forty-day period of Lent, which usually fell between February and April, and all dramatic performances ceased during Passion Week.⁵⁸ Although the enforcement of such restrictions was declining by the mid-nineteenth century, they still applied in major theaters. Patent theaters, those possessing a license to perform serious spoken dramas, such as Drury Lane and Covent Garden, still followed the restrictions. Minor theaters, however, often flouted the limitations with impunity.⁵⁹ The bans were undoubtedly not welcomed by actors and theater managers, yet they presented an opportunity for others seeking an audience. As a *Theatrical Observer* notice criticized, the "stupid practice" of closing theaters for dramatic performances during Lent "ought certainly to be abolished" in this enlightened age. The ban was nevertheless a means of bringing before the public other interesting and instructive shows. To stay in business and avoid the bans, theaters sought alternative forms of entertainment during Lent for avid theatergoers. Astronomical lectures were thus a sensible substitute. A *Theatrical Observer* notice also recommends C. H. Adams's Orrery to its readers, declaring that they could not "spend an evening more profitably than by attending a lecture either on a Wednesday or a Friday."⁶⁰

The religious tone commonly expressed in popular astronomy lectures was further justification for their presence during Lent. Lenten astronomical lectures were full of educational values, encompassing scientific knowledge, moral teachings, and divine inspiration. Religious appeal was often stressed by lecturers and reviewers, whether as a genuine motive or a rationalization for the Lenten occasion. For example, in a reviewer's words, "these lectures are particularly appropriate to the season, for religious and solemn aspirations ought to be cherished, more than any other period of the year."⁶¹ Lecturers made similar points in their advertisements. R. E. Lloyd claimed that his lectures were intended to guide the inquiring mind through Nature, "up to Nature's God," so "Seminaries and Families will find the present offer peculiarly interesting and grateful."⁶² To put it bluntly, Charles Popham even used "APPROPRIATE TO LENT" as the title of the advertisement for his lecture at Theatre Royal, Leicester.⁶³

Lectures Inside Scientific Institutions

By the early nineteenth century, London had grown into one of the largest cities in the world. The prosperity of the city's economy not only helped the cultural and entertainment industries to flourish but also made the metropolis a gathering place for people who were fond of science. Many scientific institutions and learned societies were springing up during this period; these newly emerging organizations became important forces for promoting science pursuits and public scientific lecturing. In London, scientific institutions that aimed at a broad public burgeoned at the turn of the nineteenth century and in its first two decades; the most notable of these were the Royal Institution (founded in 1799) and the London Institution (1805).[64] In addition to these establishments open to the public, learned societies also functioned as specialists' clubs where members could discuss their common interests. The Linnean Society (1788), the Geological Society (1807), and the Royal Astronomical Society (1820) were some of the most prominent examples. Outside the metropolis, similar literary and philosophical establishments also grew fast and became centers of local intellectual life, but they were hardly as resourceful as their metropolitan counterparts in visibility and scientific prowess.[65]

Science practitioners and enthusiasts could not resist convenient metropolitan networking and research resources. Many contemporaries expressed their love–hate relationship with about London's double-edged urban environment. As the young Charles Darwin claimed in a letter, "no place is at all equal, for aiding one in Natural History pursuits, to this odious dirty smokey town."[66] Another naturalist, Edward Forbes, who was forced to look for employment after losing his father's financial support, found an extra job as curator at the Geological Society. Forbes admitted to a friend that the opportunities of having "a good library at my disposal" and "personal communication daily with the leaders of science" were the two main reasons this job had attracted him.[67] Daniel Moore, a London solicitor and keen patron of science, also spent time enjoying metropolitan scientific resources. Moore, a bachelor, "had always resided in his chambers" in Kentish Town but nevertheless actively participated in the activities of arts and science circles. He was a fellow of several learned societies, including the Royal, Antiquarian, Linnean, Astronomical, and Horticultural Societies. Furthermore, the Royal Institution benefited from Moore's patronage when he generously loaned the institution £1,000 "at a time of need promptly" without interest, and eventually bequeathed the sum to the institution in his will. He was actively involved in the Royal Institution's management and became a member and later a manager of the institution's committees of general science and mechanics. Moore's reward for his patronage was the right of free access to these organizations' events. According to his obituary, "[Moore's] chief amusement was among the learned societies, where his good humour and love of science always insured a hearty welcome."[68] There was no other place like London where a science enthusiast could indulge his interests so lavishly.

So, what facilities would a scientific institution usually have in the early nineteenth century? A notice in the *Quarterly Journal of Science, Literature and the Arts* revealed clues to what science enthusiasts could expect to find in such institutes. It was an idealistic plan for a local literary and philosophical institution in the fashionable spa town of Bath, but the same expectation also could apply to similar establishments in London and other places. "It is proposed to form in Bath an Institution for the cultivation of Science, Literature, and the Liberal Arts." The notice described the scheme for the prospective scientific institution:

> The Institution to consist of a house and establishment, comprising the following accommodations: namely, a library and reading-room, from which newspapers and political pamphlets shall be excluded; a botanic garden; a museum of natural history; a cabinet of mineralogy; a cabinet of antiquities; a cabinet of coins and medals; a hall for lectures, with suitable apparatus for the courses on chemistry and the several branches of natural philosophy.
>
> To these will be added an exhibition gallery, for the reception and display of paintings and other works of the fine arts.[69]

Some of the facilities included in this prospectus were essential among contemporary scientific institutions, such as the library, reading room, and lecture hall. Others, like the art gallery and collections of curiosities, might be deemed luxuries that were not necessary in general. The facilities of a scientific institution varied depending on managers' visions, purposes, finances, and public needs. Nevertheless, exhibiting scientific knowledge and nurturing intellectual activities were the common grounds for institutions of the kind. Organizing public lectures and maintaining a library for members were basic services provided by an institution.

Furthermore, some research-orientated institutions, such as the

Royal and London Institutions, hosted laboratories and hired staff to conduct original experiments. They were important forums for scientific practitioners to make themselves known to the public. Humphry Davy, Michael Faraday, and John Tyndall all rose to fame through professorships at the Royal Institution. For wider institutional members or subscribers, an institution offered many requisites of a good club—a place to listen to the latest discoveries, gossip, or intellectual subjects, and opportunities to participate in social occasions where they could acquaint themselves with men of status in various fields.[70] This social function was probably more important than scientific instruction for many attendees in institutional events.

The Royal Institution, in particular, was a role model for providing institutional lectures to the public in the wake of the industrial age. The institution's founders envisaged propagating and applying useful knowledge in different branches of practical science like chemistry and agriculture. They proposed to build an "Institution For diffusing the Knowledge, and facilitating the general Introduction, of Useful Mechanical Inventions and Improvements; and for teaching, by Courses of Philosophical Lectures and Experiments, the application of Science to the common Purposes of Life."[71] This visionary proposal was soon taking shape on Albemarle Street in Mayfair, the site at which the original institution remains today. Despite the shadow of continuous financial crisis in the early years, the Royal Institution proved to be successful in its goal of diffusing practical knowledge, as was evident in its inspiration of many imitators modeled on the institution's scheme. The most notable imitators in the metropolis included the London Institution in the City of London, the Surrey Institution in Southwark (1807), and the Russell Institution in Bloomsbury (1808). Together with the Royal Institution, they formed the "big four" in the metropolitan marketplace of scientific institutions.[72] The Surrey Institution, however, did not last long and was dissolved in 1823. Its rotunda was later used for various purposes as a theater, a music hall, a diorama exhibition, and even for a brief time in the early 1830s as a center for radical lectures aimed at radical political reformers rather than the wealthier middle-class audience for which the Surrey Institution was intended.[73] The competition between the other three lasted much longer: the Russell Institution survived until the late nineteenth century, and the London Institution closed in 1912. The London Institution, backed by wealthy supporters from the City of London, was a considerable rival to its Mayfair counterpart. The fund that the London Institution accumulated within one year of its establishment was fourfold that of the Royal Institution.[74] Its staff was certainly aware of the rivalry between the institutions, and sometimes the competition became bitter. Faraday, for example, had once responded angrily to doubt about the Royal Institution's preeminence voiced by William Upcott of the London Institution. He felt "amused and a little offended" by Upcott and asserted that "to the world an hour's existence of our Institution is worth a years [sic] of the London."[75]

Astronomy was a well-developed lecturing subject at the Royal Institution, though the institution focused more on other subjects such as chemistry and mechanics. The earliest astronomical lecture at the Royal Institution was delivered by Thomas Young, who was appointed as a professor of natural philosophy in July 1801. Young organized a series of natural philosophy courses including astronomy during his short stay at the Royal Institution, and the six astronomical lectures he delivered between January and February 1802 were recorded in the first volume of the institution's journal.[76] The style of Young's lecturing was, however, tedious and abstruse. On many occasions he just read some passages or extracts from other scholars' papers, such as Robert Hooke's experiment to prove the motion of the Earth and the nature of gravitation, and Giuseppe Piazzi's letters to the Royal Society on the discovery of the planet Ceres.[77] Compared to his colleague Humphry Davy, Young was an ineffective lecturer and was unwilling to entertain the audience. He often misjudged an audience's ability to apprehend the content of the lectures. Though Young undoubtedly made considerable efforts to prepare the curriculum, and reviewers praised his later publication of the lectures, he failed to deliver attractive speeches.[78] By all accounts Young's natural philosophy discourses were poorly received and his astronomical lectures were probably no exception.

A few courses on astronomy were delivered at the Royal Institution after Young's departure. Table 2.1 lists courses of lectures on astronomy or relevant subjects at the Royal Institution before 1862. Compared to chemistry, astronomy was not the focus of attention in the curriculum of the Royal Institution. A course of lectures on natural philosophy often covered more content on mechanics and technical subjects such as the steam engine. Throughout the nineteenth century,

TABLE 2.1. ASTRONOMICAL COURSES AT THE ROYAL INSTITUTION

Year	Lecturer	Title/subject	Type*
1801	Thomas Young	Astronomy	Course (6)
1809	John Dalton	Natural Philosophy (included astronomy)	Course
1809	John Pond	Physical Astronomy and Its Applications	Course (10)
1810	John Pond	Popular Astronomy	Course (10)
1815	Charles Babbage	Astronomy	Course
1826	John Wallis	Astronomy	Course
1826	John Wallis	Astronomy	Juvenile
1838	John Wallis	Astronomy	Juvenile (6)
1846	John Wallis	The Rudiments of Astronomy	Juvenile (6)
1850	Baden Powell	Astronomy	Course (8)
1851	Baden Powell	Cosmical Philosophy	Course (7)

Lectures on astronomical subjects at the Royal Institution in the early and mid-nineteenth century. This list includes regular courses and juvenile lectures (Christmas Lectures).

* The numbers in parentheses indicate how many lectures were included in the curriculum if available.

Sources: Royal Institution: GB, vol. 1: 63, 139; GB, vol. 2: 47, 56, 61; *Journals of the Royal Institution of Great Britain* 1 (1802): 86–89, 108–9; *Quarterly Journal of Science, Literature and the Arts* 21 (1826): 114–16.

most of the lecturers who had given astronomical lectures at the Royal Institution were external. They nevertheless constituted a select group of prestigious men of science. These speakers included the chemist John Dalton, a member of the Manchester Literary and Philosophical Society who later became a fellow of the Royal Society; the astronomer John Pond, a fellow of the Royal Society who later became Astronomer Royal; and Baden Powell (1796–1860), the Savilian Professor of Geometry at Oxford. Among them, Powell was the most actively involved in public lecturing at the Royal Institution, yet he might be relatively unfamiliar to today's readers and researchers of the history of astronomy—he is far less famous than his son, Robert Baden-Powell, the founder of the worldwide Scout Movement.[79] As a theologian and an ordained priest of the Church of England, Powell was nevertheless keen to participate in the activities of metropolitan scientific circles and was a leading contributor to the Royal Astronomical Society and

GEOGRAPHY

TABLE 2.2. FRIDAY DISCOURSES ON ASTRONOMICAL SUBJECTS AT THE ROYAL INSTITUTION

Year	Lecturer	Title	Attendance
1850	Baden Powell	Optical Phenomena in Astronomy	
1851	Baden Powell	The Rotation of the Earth	
1851	G. B. Airy	The Total Solar Eclipse of 1851	610
1853	G. B. Airy	The Results of Recent Calculations on the Eclipse of Thales and Other Eclipses Connected with It	495
1855	G. B. Airy	The Pendulum-experiments lately made in the Harton Colliery, for ascertaining the mean Density of the Earth	405
1858	Baden Powell	On Rotatory Stability; and its Applications to Astronomical Observations on board Ships	422
1858	Charles Piazzi Smyth	Account of the Astronomical Experiment on the Peak of Teneriffe in 1856, illustrated by Photographs.	521
1859	J. H. Gladstone	The Colours of Shooting Stars and Meteors	
1861	Michael Faraday	On Mr. Warren de la Rue's Photographic Eclipse Results	783

The Friday Evening Discourses on astronomical subjects at the Royal Institution in the mid-nineteenth century.

Sources: Royal Institution: GB, vol. 2: 55, 76, 89, 104; LE, vol. 4: 48, 80; *Illustrated London News* 18, no. 485 (May 17, 1851), 419.

the British Association for the Advancement of Science.[80] He had been invited to give two courses on astronomical science at the Royal Institution in 1850 and 1851, respectively.

An exception in the Royal Institution's list of course lecturers was the young Charles Babbage, then twenty-four years old, who had just graduated from Cambridge and lacked scientific fame and lecturing experience when he started giving astronomical lectures in 1815. Like many newcomers who wanted to impress colleagues, Babbage also tried to get a foothold in the metropolitan science circles, and his brief stay at the Royal Institution gave him the opportunity to demonstrate his abilities and understanding of the science.[81] Although his attempt

FIGURE 2.8. Baden Powell's lecture "The Rotation of the Earth" at the Royal Institution. *Illustrated London News*, May 17, 1851. This image is a cutting from a scrapbook, RAS: Add MS 88: 55d. Photographed by Hsiang-Fu Huang. By permission of the Royal Astronomical Society, London.

to find a scientific position in this period was not successful, Babbage was elected as a fellow of the Royal Society in 1816, soon after his lectures at the Royal Institution.

Except for a set of courses comprising several lectures, the Royal Institution also organized one-off talks for the famous series Friday Evening Discourses. The Friday series was officially established in 1825 and soon proved to be an important initiative of the institution. Each Friday discourse generally consisted of a talk in the lecture theater followed by a soiree-style exhibition of collections in the library, which was often not relevant to the topics of the lecture. The format of a stand-alone lecture allowed speakers to talk about more specialized topics and even some current affairs of science. Table 2.2 lists the Friday discourses on astronomical subjects until 1862. Lecturers who had been invited to this series included George Airy and Charles Piazzi Smyth, the two leading astronomers who were then in office as Astronomer Royal and Astronomer Royal for Scotland, respectively. Michael Faraday had once given a talk about the photographic observation of a solar eclipse. Baden Powell had also been invited to the Friday discourses three times in 1850, 1851, and 1858. One lecture, on the rotation of the Earth, had been reported by the *Illustrated London News*, with an illustration of Powell's demonstration of the Foucault pendulum (fig. 2.8).[82]

This list of celebrities shows the prestige of the Royal Institution and its roots in the social and scientific elites. Members, subscribers, and guests of the Friday discourses expected to hear cutting-edge knowledge from the most authoritative scientific minds.

Not all astronomical lecturers active in scientific institutions had academic positions or learned society fellowships. Some freelance lecturers moved from one institution to another and were hired on a temporary basis without a permanent appointment. They ran their lecturing businesses as the private lecturers did, only with the difference that they did not lecture in a theater or an entertainment venue but preferred a scientific institution. One significant yet obscure example was John Wallis (1788–1852).[83] Wallis frequently delivered lectures at the Royal Institution: he gave a course to an adult audience in 1826 and three series of Christmas juvenile lectures in 1826, 1838, and 1846.[84] Despite his recurrent lecturing at the Royal Institution, little is known about Wallis's life and education. He was a London resident and, according to the *Gentleman's Magazine* obituary, died aged sixty-five at Camberwell on December 12, 1852.[85] His health had already deteriorated one year before his death and the illness forced him to cease all lecturing engagements. A newspaper notice in December 1851 states that the lecturer canceled his Christmas visit to Oxford due to "illness and domestic affliction."[86] In a letter of April 1852 from Wallis to E. W. Brayley, the librarian and lecturer of the London Institution, Wallis also admitted that his "worn out constitution" would not allow him to continue lecturing.[87] Although the veteran lecturer's situation seems bleak in the last years of his life, Wallis was in great demand throughout the 1830s and the 1840s. His lectures were evidently popular. The average attendance at Wallis's Royal Institution juvenile lectures in 1838 was 261 people, more than the 245 attendees of lectures the following year by the institution's professor of chemistry, William T.

Brande. Wallis's course at the London Mechanics' Institution in 1826 was also expected to draw many people, so the manager announced that each lecture would be "repeated on the succeeding Friday," hoping to prevent the venue from overcrowding and ensuring that the audience would "obtain a favourable view of the splendid machinery and transparencies."[88] Another indicator of Wallis's popularity was his tight schedule among different institutions. During the Christmas season in 1838, Wallis was engaged at the Royal Institution, meanwhile also giving a course at the London Institution. These lectures even occurred on the same day—the Royal Institution's in the afternoon and the London Institution's in the evening.[89] Perhaps because of his enjoyable lecturing style and his relatively modest fees, institutions were pleased to employ him.[90] Nevertheless, like many other freelance or extramural lecturers who taught extracurricular courses without a formal appointment, Wallis still had to fight for lecturing opportunities. His correspondence with secretaries of institutions shows that his lecturing proposals were not always successful.[91]

Wallis's lectures at the Royal Institution provide us with a typical example of the format of an astronomical course at a scientific institution. For instance, the Christmas juvenile lectures delivered in the 1846–47 season were composed of six lectures.[92] Each lecture addressed a theme and relevant topics, such as the doctrine of universal gravitation and planetary motion, and took place every two or three days. The six lectures formed a coherent syllabus. This structure allowed the lecturer to elaborate on a subject in detail and develop lecturing subjects step by step. In contrast, Lenten lectures in theaters were single events. They might be delivered more than once a week or even on consecutive days, but each lecture basically repeated the same program. A typical Lenten astronomical lecture lasted two or three hours and usually had intermissions for presenting luxurious orchestral or choral music. Such arrangements made Lenten lectures in theaters more like a spectacular show than a curricular discourse at a scientific institution.

Astronomy for the People

Like many other itinerant astronomical lecturers, Wallis did not confine his lecturing engagements solely to London. He regularly made lecture visits to Northern England at the pinnacle of his career during the 1830s and 1840s. Several major northern cities, including Leeds, Liverpool, and Manchester, were among the destinations of his lecture circuit. He had regular engagements at the Liverpool Mechanics' Institution, where he delivered astronomical courses almost annually from the late 1830s to the mid-1840s. A *Liverpool Mercury* article claimed that few subjects had been as often presented at the Mechanics' Institution as astronomy; furthermore, there was no doubt that the members would be glad to hear Wallis again.[93] With over three thousand members and extensive facilities including a library, a gallery, an evening school, and a day school for members' children, the Liverpool Mechanics' Institution was the largest of its kind in Britain outside London.[94] The institution employed numerous teachers and itinerant lecturers. It was certainly the desired venue for a freelance lecturer like Wallis seeking an important lecturing opportunity.

Mechanics' institutes were distinctive among the numerous sprouting establishments bearing the name "institution" in the early nineteenth century.[95] Although they shared the educational purpose of diffusing useful knowledge, mechanics' institutes and other organizations emulating the Royal Institution model in fact reached very different audiences. Despite the famous story of young Michael Faraday attending a lecture as a humble apprentice, the Royal Institution appealed mainly to middle- and upper-class audiences. The institution reflected the taste and concerns of the educated elite—if not the landed gentry, at least middle-class professionals. This was evident in the composition of the institution's governorship: the majority of the board was initially landed gentry during the first decade, but it was rapidly taken over by the medical and legal professions—namely, physicians and barristers—in the 1830s. The historian Morris Berman, examining the social and ideological backgrounds of the members of the institution's governorship, argues that a group of "Utilitarians" held a commanding role in the direction of the Royal Institution.[96] Another piece of evidence was the cost of being involved in institutional activities: the annual subscription for lectures at the Royal Institution was 5 guineas and an annual membership could reach 55 guineas in combination with other fees.[97] This charge was apparently not cheap for lower-middle-class audiences. Finally, the anecdote of the "demolished staircase" in its early architecture plan revealed the snobbery of the Royal Institution: the institution planned a separate staircase as the access for mechanics and other working-class audience members, but the managers decided at the last minute to remove this separate entrance from the plan.⁹⁸

Similar institutions, even outside of London, were not easy to join. For example, the Royal Manchester Institution was very particular about admission regulations. It ordered "Hereditary Governors" of households to pay a membership fee of 40 guineas and purchase an annual subscription for 1 guinea, which allowed them admission to all exhibitions and lectures. Anyone who met these required payments could bring their family members and "Strangers"—guests visiting their homes—to these events. Various types of tickets and passes were issued to regulate lecture attendees. These rules were sometimes quite trivial. For instance, a "stranger" must come from "a distance of twelve miles and upwards" and the governor should introduce the stranger in person by an order of his own handwriting or by registration in the "Strangers' Book" at the reception desk. The term "family" also had a specific definition, meaning "the wife or husband, the daughters and other female relatives, and *unmarried* sons and brothers under 24 years of age, so long as such parties are permanently resident in the house of the Governor."⁹⁹ This complicated system of admission on a household basis made the Royal Manchester Institution seem more like a selective club of the gentry. Not all scientific institutions had such strict and specific regulations as the Royal Manchester did, but it gives us a glimpse of the status of similar scientific establishments. To patronize or participate in one of them indicated being part of the middle- or upper-class elite.

By contrast, mechanics' institutes aimed at the lower-middle or working classes and thus offered relatively inexpensive admission fees. For instance, an annual subscription to the London Mechanics' Institution cost 24 shillings, around one-fifth of the Royal Institution's fee. There was also a more flexible option of 6 shillings for a quarterly subscription.¹⁰⁰ The subscription's many benefits included admission to the lectures every Wednesday and Friday evening, use of the reading room (open daily from 10:00 a.m. to 10:00 p.m.), and access to the library, plus the opportunity to enroll in courses in arithmetic, mathematics, drawing, geography, writing, English, and French. Admission charges could be much cheaper to institutions in rural areas, although the services might not be as luxurious as those in cities. The mechanics' institute in the small market town of Alnwick in Northumberland, for example, charged only 2 shillings quarterly and 8 shillings for an annual subscription.[101]

The starry heavens above remained a subject worth considering for artisans and the working-class audience. Although the sublime science seems far from practical use or everyday life on first impression, many mechanics' institutes arranged lectures on astronomy. Other organizations attached to or derived from mechanics' institutes, such as day schools for working-class children, also put astronomy in the curriculum. William Cooke Taylor, the Irish writer who made acute observations on the industrial north English cities and factory life, once proposed a curriculum for his ideal polytechnic school. Taylor included astronomy in his educational plan, asserting that geography and astronomy were "among the most important of the mixed sciences." By "mixed sciences" Taylor meant subjects between the pure sciences and the applied sciences: the former were related to mathematical subjects such as algebra and geometry, and the latter had practical use and were "most wanted in actual life," such as chemistry and natural history. He imagined that astronomy should be "made the theme of simple lectures," with the help of a celestial globe and an orrery, which could sufficiently "supply all the apparatus necessary for the purpose, and are a more legible book than any yet printed."[102] Taylor's plan shows that astronomy had become a part of education not only for gentlemen and ladies in the polite Enlightenment culture but also for the masses in the industrial era. John Wallis's engagement in the Liverpool Mechanics' Institution was thus not an isolated example.

Numerous lecturers delivered discourses on astronomy for working-class audiences; many of their activities could only be sparsely traced in local advertisements or newspapers. A Mr. Watson, for instance, delivered a lecture on the solar system at the Chelmsford Mechanics' Institution on a Wednesday evening in April 1837, illustrating his talk by the use of magic lantern phantasmagoria.[103] In Preston, Moses Holden, a prominent astronomical lecturer and a member of the BAAS, had close connections with the Institution for the Diffusion of Knowledge, the local equivalent of a mechanics' institute. In Sheffield, several lecturers, such as the renowned Glasgow professor John Pringle Nichol (1804–59), had presented astronomical courses at the local mechanics' institute.[104] In Birmingham, Rev. J. B. Podmore of Jesus

College at Cambridge gave a lecture on astronomy in the town hall on June 9, 1857. Podmore then published the lecture for the benefit of the Ashted Working Men's Association and for improvements to the St. James school.[105] The list could go on.

Cheap lectures for mass audiences came along with the prevalence of cheap books and periodicals. These different forms of media shared the common purpose of creating affordable food for thought for the public, especially growing working-class audiences. Whig movements of political and social reforms and the establishment of organizations such as the SDUK and mechanics' institutes were important forces propelling such progress. Improvements in steam-powered printing technology also made the mass production of knowledge possible. Thousands of cheap printed materials—handbills, pamphlets, newspapers, and magazines—were produced every year and the quantity of publications began skyrocketing. During the first decade of the nineteenth century, about two thousand book titles were published in Britain per annum; the number soared to six thousand in the 1840s, and eight thousand in the next decade.[106] Cheap periodicals such as the *Penny Magazine* (weekly; priced at 1d.) and the *Magazine of Science* (weekly; priced at 1½ d.) contained articles of useful knowledge with extensive illustrations aimed at a very wide readership ranging from the lower- to the middle classes.[107] The flourishing of the publishing market and lecturing businesses represented two sides of the same coin. Lecturers were often authors, who published their syllabuses or discourses; lecturers were also readers, who studied, imitated, and even copied other sources on similar subjects.

Knowledge was now a readily available commodity—sometimes even offered for free. In the *Poor Man's Guardian*, a penny weekly newspaper published in London, advertisements for cheap lectures on astronomy, and some could be found that were free or cost just a penny for admission, such as one organized by W. D. Saull in 1832, presumably the radical philanthropist and geological collector William Devonshire Saull. Saull delivered lectures on astronomy, geometry, and chemistry at Chapel Court near the high street at Borough every Friday evening. Although there is no account specifying the content of Saull's astronomical lecture, judging by his interest in the theory of evolution, his talk likely contained mentions of the radical science. The advertisement also remarked on the charitable cause of the lecture, with "the overplus to be given to the Victims."[108] But who were the victims? It did not say. Another advertisement in December 1833 showed that the "Society for Scientific, Useful, and Literary Information" had free lectures on machinery and astronomy at the Bowling Square Chapel. The astronomical talk was presented by a Mr. Davenport.[109] These advertisements were surrounded by other notices of low-cost publications and meetings of working-class unions. However, apart from these scarce and brief advertisements, we know very little about the details of these cheap lectures, let alone their causes, politics, and financial sponsors.

Compared with the above lectures for which we have only sporadic advertising information, we know much more about Rev. J. B. Podmore's charity lecture at Birmingham Town Hall, since the lecture content was later published. In the preface to the tract, Podmore thanked his patrons for encouraging him to publish it and acknowledged that the lecture was for the benefit of improvements to the St. James school. The renovation of the school, Podmore said, had been completed at a cost of £500, of which £300 had been paid (presumably the income from the lecture contributed a good deal to this sum). He expected the circulation of his booklet and patrons' extra donations to make up the remaining £200 for supporting the school and the local Working Men's Association.[110]

As cheap lectures became prevalent, the quality of the talks could not be ensured. The adjective "cheap," like "popular," is a polysemous word. Cheapness could refer to low price as well as low quality. Contemporaries were amazed by the variety of cheap publications and lectures; in the meantime, they were also dazzled by and even doubtful about the sea of information. A reader Mr. W. Bloor, for example, wrote to the editor of the *Mirror* to point out some errata in previous issues. Bloor claimed to be an experienced swimmer who had conducted experiments to examine the effects of pressure on human bodies underwater. In his letter to the editor, Bloor complained about some incorrect information he found in other publications:

> There fell into my hands a little publication called Instruction for Swimming; the author of which, I suppose, chose not to put his name, for a very good reason. He asserts that those who dive for any thing in water must go in with their eyes open, for when under water they cannot open them,

nor shut them when they are open. About the same time another and similar work met my eye ... the author called [it] "Doctor Franklin's Advice to Bathers;" this piece contained the same assertion. I looked [at] one of these catchpenny things through, and found such a variety of wonderful antics taught to be performed in the water, that I never saw performed or heard of, and believe no man ever did perform.[111]

Although Bloor's criticism was not related to astronomy, it still shows how easy it was to disseminate knowledge—whether orthodox or heretical, sound or unreasonable—in the age of industrial printing and publishing. Drowning in a sea of information is not an exclusive privilege of modern-day Internet users; nineteenth-century readers also got lost in the waves of cheap publications that James Secord describes as "an army of Lilliputians."[112]

Another example, this time regarding astronomical lecturing, was an anecdote in the *London Saturday Journal*. The anonymous journalist, with "some spurning feeling of contempt," read a statement from a gentleman. This assertive gentleman informed the public that he was prepared to give a lecture on astronomy, "on the principle of the *earth being at rest*; and offering his services to mechanics' institutions and scientific associations." The journalist then remarked ruthlessly: "What! said we, does this feeble body think that he can pull an 'enlightened' public back two centuries and a half?—a dwarf holding up his finger to wrestle with the giants who have scaled the heavens!"[113] This gentleman who firmly believed in and promoted the geocentric model was probably a Muggletonian rather than an uneducated country yokel. Muggletonians were a small group of dissenting Protestants who followed the teachings of two seventeenth-century plebian religious thinkers, John Reeve and Lodowicke Muggleton. In principle, Muggletonians had many unorthodox beliefs, including that God had directly appeared on the Earth as a glorified man, Jesus Christ; God takes no notice of everyday doings in the world; and the soul dies with the body and will be raised with it. They also maintained a general hostility to reason, asserting it as a trick born of desire and lack that will misguide humans into taking whatever they want. Therefore, Muggletonians rejected many orthodox opinions of theology and philosophy, including Newtonian cosmology. One representative Muggletonian author in early Victorian England was Isaac Frost, who penned the treatise *Two Systems of Astronomy* (1846) to elaborate a scriptural understanding of the universe and to discredit Newtonian astronomy.[114] We are not certain whether the gentleman in the journal report was Isaac Frost, or if he actually gave any talk on his astronomical findings at any institution. This anecdote nevertheless suggests that the quality of public lectures was uneven in an age when anyone could publish or talk about anything.

Every Nook and Cranny

We have visited various sites of astronomical lectures ranging from theaters to mechanics' institutes. Our journey began in London and we explored places outside the metropolis, such as Liverpool, Manchester, and other provincial towns. Theaters, scientific organizations, and mechanics' institutes were familiar venues located in these cities and towns for astronomy lecturing. The countryside offered no such venues, yet the heavens still appealed to curious minds in rural areas. Itinerant lecturers traveled through towns and the countryside with their lecturing gear, reaching every nook and cranny of the British Isles.

In a place without a theater or a lecture hall, lecturers might flexibly arrange astronomical talks in a "temporary" venue that normally functioned as a gathering place for local people. Town halls, assembly rooms, schools, libraries, churches, patrons' houses, and even pubs were all potential venues to be borrowed for a lecture occasion. It is not difficult to find these various venues in advertisements or stories in local newspapers. Some lectures were given by familiar names. The Walkers, for example, were based in London but regularly traveled to many places across Great Britain. Even lecturing in the vicinity of London, the Walkers might choose a venue other than a theater. For instance, when D. F. Walker visited Wandsworth Town—now a district of south London but then a part of the county of Surrey—in 1820, he lectured in the assembly rooms of a local pub, Spread Eagle.[115] Another artisan lecturer, Robert Children, was also a keen traveler on the lecture circuit, whose footsteps reached the town halls at Oxford in November 1835 and at Sudbury, Suffolk, in April 1841. He also delivered two lectures at Mr. Hellyer's Library in the seaside town of Ryde, Isle of Wight, under the patronage of the Rev. W. S. Phillips in 1844.[116] A Mr. Goodacre presented astronomical lectures in a riding school at Huddersfield in August 1822. After the lecture's conclusion, Mr. Goodacre announced that his next stop would be nearby Halifax, where he had solicited public patronage.[117] These are just the slightest traces of some itinerant lecturers' activities.

Outside Great Britain, astronomical lectures were also given in other lands of the United Kingdom or the Crown Dependencies, even on the remote Channel Islands. The English genealogist William Berry, for example, lectured on astronomy with the transparent orrery in Guernsey. The Priaulx Library at Guernsey preserves Berry's papers including a lecture poster dated 1811.[118] Turning our attention farther west—in Ireland, the Rev. William P. Moore, the headmaster of the Royal School at Cavan, delivered a lecture titled "The Wonders of Astronomy" before the Belturbet Literary and Scientific Society between 1854 and 1855. Probably because of its educational value, Moore's lecture was somehow included in the lecture collection organized and edited by the Young Men's Christian Association in Dublin, despite not having taken place in the common series venue of the Rotundo in Dublin.[119]

Sometimes astronomical lectures in rural places more resembled an outreach service or impromptu invitation by local patrons than the commercial event of an itinerant lecturer. For instance, a series of lectures on astronomy was presented at the Rev. M'Alister's meeting house in Holywood, Ireland, by a schoolteacher, Mr. MacKeown of Belfast, in November 1842.[120] Dr. Bateman, who stayed in the spa town of Ilkley for a holiday during May 1852, was invited to give a gratuitous lecture on the astronomical articles displayed in the Great Exhibition at Dr. Macleod's local hospital.[121] Under the patronage of the Illford Mutual Improvement Society, a schoolmaster, Mr. Joseph Freeman, delivered an astronomical lecture at the Baptist Chapel School, Great Ilford, Essex, in January 1856.[122] In Harwich, Essex, a lecture on "practical astronomy," delivered by a Captain Campbell of Her Majesty's Indian Navy, was held at the assembly rooms kindly lent by R. J. Bagshaw in March 1866. The journalist claimed that the "gallant lecturer" was "frequently and deservedly applauded by a large audience." The proceeds of the lecture, about £5, were donated to the funds of the Shipwrecked Fishermen and Mariners' Benevolent Society.[123] We know very few details of these events other than the date, place, and lecturer's name. Nevertheless, these bits of information show the prevalence of astronomical lecturing even in the countryside.

Most of the examples listed above took place in a small provincial town or rural settlement. The populations of these places, except for Oxford, Huddersfield, and the vicinity of London, consisted of about ten thousand or less in the mid-nineteenth century.[124] There was hardly any theater or scientific institution in the neighborhood of these rural parishes. Patronage of the local gentry and savants was often important for arranging a guest lecturer to pay a visit. Clergy and schoolmasters were especially willing to support or actively organize such lecture occasions because of their educational value for local audiences. Schools and churches were thus common venues for itinerant astronomical lectures.

As the *Chambers's Edinburgh Journal* article mentioned at the beginning of this chapter indicates, the arrival of an itinerant astronomical lecture in a smaller provincial town was usually an "event productive of general satisfaction" that served to "enliven one or two of the dreary weeks of winter."[125] There was a shortage of entertainments in small provincial towns and the countryside, especially for the working classes. Astronomical lectures were often accompanied by displays of orreries and slideshows. These visual appliances entertained nineteenth-century audiences as much as films did half a century later. Therefore, a visit from an itinerant lecturer was a sensation for locals during humdrum days. This crowd-drawing power was even more significant when the visitor was a scientific celebrity, like Astronomer Royal George Airy. When Airy visited Ipswich to deliver a course of six astronomical lectures in March 1848, around 700 locals attended—considering that the population of Ipswich in 1851 was not over 27,000, this attendance was impressive.[126]

Apart from entertainment appeal, itinerant lectures had immense educational value, especially for rural areas that lacked resources. A short but touching story in a local newspaper illustrates such an opportunity for country pupils. When a Mr. T. P. Barkas gave a lecture on astronomy at the Primitive Methodist Chapel, Earsdon, in March 1852, he offered the prize of a book at the conclusion of his lecture. This "neat and useful book" was presented to a youngster "who answered correctly the largest number of questions asked by the lecturer relative to astronomy."[127] Earsdon was a township, parish, and subdistrict in Tynemouth, Northumberland, in the far northeast of England. The population of Earsdon was between 8,400 and 11,000 in 1851, about

60 percent of whom were unskilled workers and servants.[128] At a small rural parish far from cities, receiving a proper education was a luxury for local children. To give a lecture and a book to a youngster was thus like sowing and watering seeds in the soil.

In recent decades, geographers and historians of science have searched high and low for different sites and places where scientific knowledge was produced, circulated, reacted to, and transmuted. Their research shows that science can happen in a multitude of unexpected localities and prompts us to think about the influence of spatial factors on scientific thought and practice. Bernard Lightman, for example, has surveyed various science sites that produced elite epistemes in nineteenth-century London, including strongholds of gentlemanly science, utilitarian science, and scientific naturalism, as well as sites that resisted the above ideologies. Lightman shows how these sites were reconfigured and refashioned to serve the needs of various groups or individuals through time. For instance, Kew Gardens was transformed from a space of landed aristocracy and the upper classes that emphasized the importance of agricultural science, to a national public institution illustrative of Britain's economic and imperial power by committing to pure botanical research.[129] Lightman also extended his reach to sites where not-so-elite epistemes of science were exhibited, such as the Colosseum in Regent's Park. He asserts that the question now no longer involves proving that places mattered, but *how* space mattered.[130] In line with this approach, Diarmid Finnegan's study of scientific lectures in mid-Victorian Edinburgh emphasizes the dynamic entity of speech spaces that emerged from the interaction between local protocols and cultures, rather than the mere locations where these lectures were held. In Finnegan's analysis, the effects of the spaces of lecture theaters and the voice of speeches both shaped the performance of the lecture and determined whether it was articulate and impactful.[131]

Sites of nineteenth-century astronomy, too, were not limited to observatories and the relevant learned societies that cultivated gentlemanly science and an elite episteme. This chapter has explored a myriad of venues in which popular astronomy lectures took place during the first half of the nineteenth century. Many of these venues were unconventional, such as theaters in entertainment districts, assembly rooms in local pubs, and churches and country estates in rural areas. The spatial characteristics of these venues certainly shaped very different kinds of performance for astronomy distinct from those that took place in scientific institutions. These venues were not sites where astronomical knowledge was originally produced, but they were places where astronomical knowledge was in transit, to be transformed and multiplied.

Among the miscellaneous sites of popular astronomy, theaters were particularly important in the first half of the nineteenth century. This is due not only to the theaters' potential for hosting large audiences but also to the facilities' capabilities for contributing to the rise of the genre of stage astronomy. The theatrical turn initiated by Adam and William Walker from the late eighteenth century onward expanded audiences and made popular astronomy lectures in this period a unique performance compared to other scientific lectures. Stage astronomy in theaters became the embodiment of the sublime science. Other popularizers of astronomy felt the impact of this kind of performance and motivated them respond, whether they loved it or not.

CHAPTER 3

AFFILIATION

The nineteenth century witnessed a drastic transformation of the scientific community. At the beginning of the century, the practice of "gentlemanly science" still prevailed in Britain. Most scientific researchers were gentlemanly specialists devoted to career pursuits of science but did not depend on them for a living.[1] This practice clearly differed from paid professional work for governmental, industrial, or academic institutions toward the end of the century. Today scientific communities have largely consolidated into professional organizations. Although the portrayal of a lone, often eccentric genius working in a basement or garage laboratory is common in the mass media, real-world science does not work this way. An independent scientist without any institutional affiliation is rarely, if ever encountered. The roots of professional scientific institutions are often traced to the beginning of the nineteenth century, when the institutional sites of analysis and professional education started developing.[2] Nevertheless, this transformation did not occur suddenly. The rise of scientific professionals did not bring an abrupt disappearance of amateur practitioners.[3] The situation in the scientific lecturing trade was similar, for the rise of institution-affiliated lecturers did not lead to the decisive withdrawal of independent competitors.

Who were the lecturers of astronomy in early Victorian Britain and before? What were their identities and institutional affiliations? The popular astronomy lecturing trade remained a shared arena of private and institutional practitioners in the first half of the nineteenth century. The rivalry between the two popularizers who were active in early

AFFILIATION

Victorian London, C. H. Adams and George Bachhoffner (1810–79), exemplifies this pattern. The boundaries between institutional men of science and private entrepreneurs were blurred in the popular astronomy lecturing trade throughout the first half of the nineteenth century. Some itinerant lecturers, such as John Wallis, wandered around in the gray area. Wallis was regularly employed by scientific institutions but he was not appointed to any permanent post by a particular institution. For a long time, he seemed to do freelance work. The association of itinerant lecturers like Wallis with scientific institutions is therefore disputable.

Our quest for information on astronomical lecturers' identity and affiliation inevitably leads to another question about how they built a lecturing career. Although "scientist" (a term that was not coined until 1834) as a profession in the present-day sense did not exist in the early nineteenth century, prospects of science as a career had already emerged.[4] Historians have looked into the development of career pursuits in different disciplines of science in and before the nineteenth century. For example, Roy Porter focuses on the career of geologists: "those who spent a significant amount of their lives doing geology." Porter elaborates almost exclusively on researchers rather than other groups of practitioners.[5] In the realm of astronomy, Allan Chapman has explored the career and practices of Victorian independent astronomers in detail.[6] Like Porter, Chapman focuses on researchers, despite mentioning a few cases of popular lecturers as a part of broader independent practices of astronomy. Prospects of astronomy lecturing as a career had also emerged in Britain before the Victorian era. As I have indicated in chapter 1, the astronomy lecturing trade was actually an extension of public natural philosophy lecturing in the eighteenth century. The emergence of mass media in the industrial era, plus a growing need for education and entertainment, provided lecturers with a larger market and audience.

The notion of career here, however, should not be considered narrowly in mere professional terms. While the meanings of "career" and "profession" often overlap, they are not the same. The notion of a career in the nineteenth century was more vocational (for a lifetime) than occupational (to make a living); the *Oxford English Dictionary* defines "career" as "a person's course of progress through life (or a distinct portion of life)."[7] Porter also reminds us that the development of the concept "career interests" during the Victorian era contributed to contemporaries' distinction between a leisured amateur and a career practitioner.[8]

This chapter starts with a description and comparison of the careers of C. H. Adams and George Bachhoffner. These two lecturers are relatively obscure for modern-day readers. Adams and Bachhoffner represent examples of the private and institutional lecturer, respectively. In their analysis of public scientific and astronomical lecturing in Britain before 1850, both J. N. Hays and Ian Inkster indicate that a decisive institutionalization of scientific lecturing occurred in the early nineteenth century.[9] The comparative cases of Adams and Bachhoffner, however, refute Hays's and Inkster's assertions. My analysis based on a categorization of private and institutional lecturers is nevertheless not without problems. As Bernard Lightman warns in his study of Victorian popularizers of science, the demarcation between popularizers and practitioners should not be seen as overly rigid.[10] The dichotomy between institutional and private lecturers is also porous and should be treated with caution. This ambiguity also applied to qualifying as a popular astronomy lecturer. A lecturer's credentials were not necessarily based on his educational background or involvement in astronomical research. I use the cases of two "artisan" lecturers, John Bird (d. 1840) and Robert Children, to explore the financial and social reasons that private lecturers engaged in the lecturing business. Making profits and raising social status from public lecturing were important motivations. The concept of self-improvement played important moral and economic roles in nineteenth-century astronomy lecturing.

C. H. Adams, "The Only Orthodox Interpreter"

To early Victorian Britons, Lent was the astronomy season. Astronomical lectures were often arranged as an alternative fixture during the forty-day religious vigil, when theaters were closed and regular plays were banned. Newspapers and theatrical journals would report on the annual Easter entertainments for those pleasure seekers who did not follow the ascetic teachings of the Church. One story in the *Illustrated London News* on April 11, 1857, for example, summarized the events of that season in the metropolis. The shows listed were of all kinds: an exhibition of dissolving views, a comic ballet, a German wizard's magic performance, a narrative of a lion-hunting adventure, and many

AFFILIATION

more. Among the multifarious entertainments, two separate celestial exhibitions were presented: "Mr. C. H. Adams has, as usual, presented his Orrery at the HAYMARKET during the week, accompanied with his annual lecture on astronomy... while Dr. Bachhoffner has delivered at the COLOSSEUM a daily lecture on astronomy, illustrated by a new and beautiful orrery."[11] These two exhibitions of heavenly bodies were on a collision course in the arena of Lent amusements.

C. H. Adams and Bachhoffner were two shining stars of the popular astronomy lecturing trade in early Victorian London. They represented two different modes of business operation: Adams, a private showman, had no links with the scientific establishment, while for the most part, but not always, Bachhoffner's lecturing career was affiliated with a particular scientific organization, the Royal Polytechnic Institution. Distinctions between the career patterns of these two lecturers offer us specific means to analyze the identity of the practitioners in the marketplace.

We start with C. H. Adams, whom a critic had given the title of the "only orthodox interpreter" of astronomy. Adams's celebrity status among mid-Victorian London showmen is exemplified by an extract from *Punch's Almanack* for 1861, the special annual issue of the renowned satirical magazine *Punch*. This extract lampoons solar eclipses and relevant businesses including Adams's:

> ECLIPSES.—We are happy to inform our readers that the Astronomical Society of London has at length succeeded in rectifying the globe, and that in future there will be no more eclipses. The holes in the sun's path have been carefully filled up with concrete of diamonds, the Zodiac has been duly and completely oiled, and all the houses that were in opposition have been pulled down. The course of the planets will henceforth be regular. Compensation has been demanded by about eleven thousand street boys, who were in the habit of selling smoked glasses to view the old phenomena, and the claimants have been sent to the Compulsory College, and Mr. Adams, the lecturer, who on Saturday attained his six hundredth year, has put fireworks instead of eclipses into his famous Orrery, with which our young folks are much better pleased.[12]

The original piece in *Punch* had a futuristic subtitle, "From Punch's Almanack for 2417." The lecturer Mr. Adams would have attained his "six hundredth year" in this imaginary future. In fact, at the moment this extract was published, C. H. Adams had already completed thirty consecutive years of London performance and was in the middle of his thirty-first year. The show's longevity in the metropolis was evidence of its commercial success. It was also a significant achievement because few private lecturers could sustain their businesses for such a long period. The fact that Adams and his show became targets of *Punch*'s ridicule just reflects the lecturer's fame as a household name.

Adams was perfectly aware of the advantage of his consecutive performances. As a selling point to promote his show, he repeatedly boasted about its continuity. Advertisements for Adams's lecture would stress how long the show had existed, and slogans like "TWENTY-FOURTH YEAR in LONDON" were printed in bold text.[13] The longevity of Adams's lecture was not only a matter of fact but also a self-branding that helped keep Adams's business in the lead. The *Era*, one of the most influential weekly theatrical journals in Victorian London, commented in 1856 that Adams's performance was still competitive in confronting other challengers in the marketplace. The journalist asserted: "Long ago, when the present middle-aged gentleman and father of a family was a little boy . . . the lecturer [Adams] commenced his illustrations of the wonders of the starry heavens; and from that time, though Panopticons and Polytechnics have arisen in the interval, and done much to elucidate the same subject, there has yet been a firm faith kept alive in the minds of the public that Mr. Adams was the only orthodox interpreter of the phenomena attendant on the revolutions of the celestial bodies."[14]

Charles Henry Adams, the man later commonly known by audiences as C. H. Adams, was born in Edmonton, Middlesex (today a part of Greater London), on February 22, 1803, and died in the same place on November 15, 1871.[15] Very little is known about Adams's early life and educational background. Because his father was a local schoolmaster, it is reasonable to suppose that he had received an appropriate education. Adams succeeded his father as the headmaster of the Latymer School, a grammar school in his hometown. The Latymer School offered education in grammar and Latin for needy pupils between five and seventeen years old. The school also had formal links with St. John's College, Cambridge, which funded scholarships to the university for Latymer's graduates.[16] Adams earned a living in this post when not lecturing on astronomy. His wife Jane Adams (née Sawyer) was also a teacher. The couple married in 1834, and at least four of their children, two daughters and two sons, survived into adulthood.[17]

We are not certain what exactly motivated Adams the schoolmaster to step into the business of astronomy lecturing, but there were already many models for him to imitate before his engagement in stage astronomy. George Bartley, the Walkers, and many other Lenten or itinerant lecturers, all showed that it was possible to pursue a career in public lecturing. Adams had made his stage debut by 1830. A playbill in the *Theatrical Observer*, dated April 6, 1830, advertised that Mr. C. H. Adams would deliver a lecture on astronomy in the evening at the Royal Adelphi Theatre, the Strand.[18] This is the earliest known record of a lecture by Adams. Afterward, Adams developed a remarkable thirty-two-year lecturing career until 1861. An advertisement in the *Morning Chronicle*, dated March 22, 1861, was one of the last known documentations of an Adams lecture and probably was also evidence of his final season of lecturing.[19] By 1864, Adams's lecturing business had already ended. A newspaper article summarized "some changes amongst the general entertainments," including Mr. Adams, "with that horrible Orrery, no longer frighten[ing] children of tender years and women."[20] Throughout the thirty-two years, Adams delivered astronomical lectures annually in London during Lent and occasionally lectured outside London. The name of the lecturer, Mr. C. H. Adams, was well known in connection with his Orrery, as indicated in the very brief notice of his death in the literary magazine the *Athenaeum*.[21]

Despite the show's longevity, Adams's lecturing business did not always operate smoothly. The *Literary Gazette* reported an incident on March 17, 1832. The theater was vandalized in the afternoon before opening hours, forcing Adams to conduct his lecture in a rather difficult situation.[22] The newspaper reported that some evildoers had contrived to enter the auditorium, "abstract the glass of the lantern, break in the face of the orrery's sun, knot up all the cords, and commit other mischief, so as to prevent the lecture from being illustrated in the magnificent manner." Fortunately, by much exertion, "the injuries were in great measure remedied" before the start of the show, and Adams presented the lecture as usual to a "sympathising and admiring" audience. The identity and motives of the marauders are unknown. Could this have been an act of sabotage orchestrated by a business rival? Or could it merely have been a random act of vandalism? We may never know. This incident did not deal a severe blow to Adams's business. It nevertheless showed that the venture of popular astronomy lecturing was full of risks.

Adams's lectures had received generally positive receptions from the critics since his debut. Most reviewers acclaimed his lucid style of expression and his skillful use of visual aids. These abilities made his lecture accessible to lay people even when discussing specialized topics, such as the parallaxes of stars.[23] He usually organized his lectures into three parts: first, an introduction of the Earth—its shape, dimension, diurnal and annual motions, and the four seasons; second, the Moon, particularly in relation to the terrestrial phenomenon of the tide; and finally, individual planets and the entire system, with historical cosmologies from the Ptolemaic to Copernican models.[24] A *Theatrical Observer* review in 1836 remarked that Adams, assisted by "some very ingenious apparatus," rendered the details of the science of astronomy easy to comprehend, with his distinct voice and style of explanation being "perspicuous, clear and unaffected."[25] Similar approval appeared in the *Literary Gazette* in 1842, when Adams's lecture was presented throughout the week at the Adelphi to "hundreds of young and old." The journalist praised Adams as "among the most delightful and meritorious of their purveyors" of instruction and entertainment.[26]

Except for the value of lucid expression, some critics stressed the religious merits of Adams's show as going far beyond knowledgeable instruction or pure amusement. A *Metropolitan Magazine* review encouraged its readers to attend the lecture. "Not only will [Adams's] visitors gain a rapid notion of the motions of the heavenly bodies, in a most pleasing manner," the reviewer claimed, "but also they will have afforded to them the very best manner of elevating their thoughts into piety, by the contemplation of the wonderful system of the universe, and which thoughts, though always laudable, are particularly appropriate during the continuation of Lent."[27] Contemporary readers were familiar with similar moral and religious rhetoric. It rendered the virtue of studying astronomy as a beneficial means to confirm faith and inspire devotion. The Bridgewater Treatises and many other contemporary popular scientific publications employed the same religious language and themes.[28]

The audience for Adams's lecture covered a broad range of people, from the working classes to the rich, from youths to adults. This can be seen in the different tiers of admission rates. For example, at the Adelphi Theatre's 1853 season, the least expensive ticket for a lecture was 6d. for a seat in the gallery. The mid-range prices included 1s. for a seat in the pit, 2s. for one in the boxes, and 3s. to sit in the stalls. A seat in the private boxes was much more expensive, costing 10s., 6d., or 21s.[29] The same rates applied for much of Adams's long-running career. The most expensive ticket, 21s. (1 guinea), was equivalent to about the average weekly wage of a general laborer in early Victorian London—apparently this price was not affordable to the working classes.[30] The prices for the pit, the boxes, and the gallery were comparable to those for other dramatic plays in most "minor" London or provincial theaters.[31] Charles Dickens described a working-class theater at Shoreditch in 1850, where the ticket prices were 1s. for the boxes, 6d. for the pit, and 3d. for the upper gallery.[32] The prices for admission to Adams's lecture at the Adelphi compared to those in a working-class theater were certainly much more expensive. Nevertheless, the lower tiers for tickets to Adams's lecture were still affordable to the working-class audience.

Adams attempted to win over groups of young people along with their guardians. Students or children often received half-price admission to the pit and boxes; these offers were especially common in Adams's later years. Other bargains, such as family tickets admitting four people, sometimes appeared.[33] These offers were undoubtedly tempting to teachers and parents.

The Adelphi was not the only theater Adams performed in. During his thirty-two years of lectures, Adams gave talks in various West End theaters, including the King's Theatre (renamed Her Majesty's Theatre in 1837 during the reign of Queen Victoria), the Haymarket, the Lyceum, and the Princess's. Adams's lecture was praised by critics as "the best of the substituted exhibitions" during Lent, when dramatic plays in theaters were suspended.[34]

All good things end, however, and Adams's show was unexceptional. The Passion Week of 1861 was the "31st year" of Adams's lecture and probably the last.[35] Two newspapers, the *Era* and the *Standard*, reported that this was the conclusion of C. H. Adams's exceptional stage career. The *Era* claimed that the "informative astronomical lecture with which he has for the last thirty years enlightened the public . . . appears to have lost nothing of its interest."[36] The *Standard*, remarking on the show's curtain call, said, "On the termination of the lecture Mr. Adams was called forward to receive the applause of his admiring audience."[37] After his retirement from the theater stage, Adams continued as the headmaster of the Latymer School. His management of the school was, however, less successful than his celebrated stage performance.[38] The quality of education during the final years of Adams's reign was poor. An inspection of the school in 1865 indicated that the teaching standards for Latin and elementary subjects were very low. Only two-thirds of registered students attended school in the morning and less than half of them stayed for the afternoon. The funding for the Cambridge scholarship was even used for church repairs. Adams eventually retired from the headmaster position on a pension in 1868.

Over thirty years of performance on the stage had undoubtedly made Adams's lectures iconic. After the close of the show, a *Leeds Mercury* article extracted from the *Telegraph* mentioned this legacy, referring to Adams's orrery as a part of an event comparable to other amusement landmarks. The reporter expressed strong, though not totally positive, nostalgia for many popular amusements in London from earlier times:

> To the present generation, however, Saville House had a history of its own, less classic, perhaps, but not less curious. Who amongst Londoners able to date his childhood so recently as a quarter of a century ago, but must remember the marvels of the Linwood tapestry gallery? The spectacle to which little boys and girls were taken in the humdrum days when William IV. was King was not, perhaps, one of dazzling excitement. Possibly the recollection of a morning passed at Saville House ranked with the evening at Adams's Orrery, or with the afternoon at the old Adelaide Gallery, and was not to be mentioned in the same breath with the jaunt to Vauxhall or the pantomime.[39]

In a way, Adams's show was part of the collective memory among Londoners of their childhood in the early Victorian era. Like pantomimes at Christmas and the New Year, astronomical shows with a large orrery had become conventional fixtures during Lent, and Adams's was the most popular. Even when the *Theatrical Journal* reviewed John Henry Pepper's astronomical lecture in the Royal Polytechnic Institution in 1871, the reporter used Adams's as a reference. The report referred to Adams's orrery as the best Lenten exhibition and commented that

Pepper's astronomical lecture was "a reproduction of this [Adams's] once popular representation of the revolutions of heavenly bodies, on a more refine[d] and elegant scale."[40] Adams's lectures had become a standard by which to judge subsequent spectacles of the kind.

BACHHOFFNER AND THE ROYAL POLYTECHNIC

C. H. Adams was the epitome of private lecturing on astronomy in early Victorian London. Despite the popularity of his lectures, he had far from a monopoly. Bachhoffner, a significant longtime business rival of Adams, shared a few career similarities with the Edmonton schoolmaster. Both started engaging in lecturing ventures in their late twenties: Adams made his stage debut at twenty-seven; Bachhoffner was a twenty-eight-year-old when he participated in the foundation of the Royal Polytechnic Institution. Both had retired from lecturing by the early 1860s. Their astronomy lecturing activities overlapped during a long period in the metropolis, and both lecturers enjoyed comparable popularity with audiences. However, their business operations had a distinct difference: while Adams kept his venture independent of any scientific institution, Bachhoffner was affiliated mostly with the Royal Polytechnic Institution and later became the proprietor of the Colosseum. Because of his longtime affiliation with the Royal Polytechnic, it is hard to neglect the institutional influence on Bachhoffner's career, just as it is impossible to discuss Michael Faraday without mentioning the Royal Institution.

As with Adams, the life of Bachhoffner has remained obscure. The short obituary for Bachhoffner in the *Leeds Mercury* created a very brief biography: "The death is announced of Dr. G. H. Bachhoffner, who died at his house in Hammersmith on July 22nd, aged sixty-nine. He was for upwards of thirty years a popular lecturer at the Polytechnic Institution and the Royal Colosseum, on natural philosophy, chemistry, and astronomy. The Royal Polytechnic, in fact, was originated by him, a meeting being held at his house at which he suggested the scheme."[41] George Henry Bachhoffner was born in London on April 13, 1810. Little is known of Bachhoffner's early life. He studied at the University of Giessen in Germany, where he graduated with MA and PhD degrees.[42] Bachhoffner started engaging in scientific lecturing and writing very early. Before the founding of the Royal Polytechnic Institution, while in his late twenties, he published two treatises,

Chemistry as Applied to the Fine Arts (1837) and *A Popular Treatise on Voltaic Electricity and Electro-Magnetism* (1838). The former was derived from lectures he had delivered, which several eminent artists and patrons encouraged him to publish.[43] The latter, designed for lay readers, indicated the author's affiliation as "Lecturer on Chemistry to the Artists' Society." These publications show that Bachhoffner taught chemistry courses to artists in his early career.

The Royal Polytechnic Institution, opened to the public at 309 Regent Street on August 6, 1838, was a formidable new force in London's competitive science and recreation market.[44] The founding members involved in the operation of the early Royal Polytechnic were from different classes and backgrounds, and included entrepreneurs, professionals, and parliamentarians.[45] Among these members, three men played key roles: the first chairman Sir George Cayley, the former Adelaide Gallery supervisor Charles Payne, and the builder William Mountford Nurse. As a wealthy landowner and an enthusiastic aeronautical designer, Cayley was a keen patron of scientific and technological projects. He was also the sponsor of the Adelaide Gallery, the predecessor of and model for the Royal Polytechnic Institution in displaying practical science, arts, and crafts to the public. Payne, who later became the secretary of the Royal Polytechnic, was the main organizer of this planned new institution and had secured Cayley as sponsor. Nurse, a speculative builder who focused more on potential profit, contributed to the building, and furnished a large part of the capital.[46]

Bachhoffner's role in the founding of the Polytechnic Institution is unclear; we do not know if the scheme originated precisely with him as his obituary suggested. Nevertheless, it would not be surprising if Bachhoffner had such an idea. The demand for a permanent institution where new inventions and technical knowledge could be introduced to the public never ceased during the first half of the nineteenth century.[47] Whether or not Bachhoffner contributed to the founding of the Royal Polytechnic Institution, he was involved in its day-to-day operations from the beginning. He was appointed as principal of the Department of Natural and Experimental Philosophy at the Royal Polytechnic, a post he held until 1855.

As principal of the experimental department, Bachhoffner was in charge of managing and exhibiting experiments, including the spectacular "hydroelectric machine," which produced static electricity.[48]

"THE ROTATION OF THE EARTH MADE VISIBLE," AT THE POLYTECHNIC INSTITUTION, REGENT-STREET.—(SEE PRECEDING PAGE.)

FIGURE 3.1. "'The Rotation of the Earth Made Visible,' at the Polytechnic Institution, Regent-Street," *Illustrated London News*, May 3, 1851. This image is a cutting from a scrapbook, RAS: Add MS 88: 79b. Photographed by Hsiang-Fu Huang. By permission of the Royal Astronomical Society, London.

He also delivered regular lectures, especially on topics in chemistry and physics, such as one titled "Electricity, Galvanism, and Other Branches of Natural Philosophy."[49] Although astronomy was not in Bachhoffner's repertoire in his early years at the Royal Polytechnic, advertisements in 1845 publicized his lecture on astronomy during Lent.[50] Unlike Adams, who concentrated on particular topics of astronomy, Bachhoffner was always a versatile lecturer on a wide range of subjects. He could elaborate on the rotation of the Earth by demonstrating a Foucault pendulum (fig. 3.1).[51] He also spoke on highly technical topics like "electro-gilding and silvering" and "Wilkins's New

Patent Universal Electric Telegraph."[52] These practical topics accorded with the utilitarian objective of the Royal Polytechnic Institution. He also focused special attention on the juvenile audience. For example, he claimed that his lecture on the philosophy of scientific recreation in December 1851 had been "arranged expressly for the instruction and amusement of the Junior Branches visiting the Institution during the Holydays."[53]

Bachhoffner's versatile skills and familiarity with experimental instruments made him a great asset to the early Royal Polytechnic Institution. His significance for the Royal Polytechnic was similar to the roles of Francis Hauksbee and John T. Desaguliers at the Royal Society during Isaac Newton's presidency, or to those of Humphry Davy and Michael Faraday at the Royal Institution. An example of Bachhoffner's excellent execution of experiments was his work with the hydroelectric machine. This machine, designed by the Newcastle lawyer and inventor William Armstrong, was a locomotive lookalike apparatus that produced static electricity by high-pressure steam from a boiler. Armstrong collaborated with the Royal Polytechnic to transform his invention into a sensational exhibition to the public.[54] Bachhoffner assisted in the construction of the machine at the Royal Polytechnic. He also conducted a series of tests before displaying the machine before an audience. The machine, with Bachhoffner as its demonstrator, debuted on September 15, 1843. The exhibition of the hydroelectric machine received mostly approving reviews; the critics agreed that this new spectacle would become a "lion" in the gallery of the Royal Polytechnic.[55]

The rivalry between Bachhoffner and C. H. Adams for astronomy lecturing is suggested in the newspaper advertisements. During the decade between 1845 and 1855, these two lecturers provided a large quantity of popular astronomical lectures in London during Lent. Most of the advertisements for astronomical lectures in the newspapers were for their orations. Their advertisements were usually placed on the same page, even next to each other. For example, in the *Era* on March 16, 1845, Adams announced that his "accustomed ANNUAL LECTURE on ASTRONOMY, for MONDAY, March 17th, and every evening during the week (Good Friday excepted)" will be at the Adelphi Theatre. Meanwhile, in the same column, the Royal Polytechnic Institution publicized its program that included "A Series of Lectures on

Astronomy, by Professor Bachhoffner on the Mornings and Evenings of Mondays, Wednesdays, and Fridays, during Lent."[56] The two lectures clashed with each other in the Lenten amusement arena. At the time, Adams had already been presenting his show for fifteen years; in contrast, although Bachhoffner was an experienced lecturer, astronomy was a subject he had just started engaging in. Nevertheless, with the resources of the Royal Polytechnic, Bachhoffner's lecture must have posed a serious challenge to Adams's.

Newspapers also positively reviewed Bachhoffner's venture into astronomy lecturing. For example, the *Critic* praised his performance and encouraged readers to attend the lecture: "The lecturer [Bachhoffner] endeavoured to impress upon his audience the importance and pleasure derived even from a slight acquaintance with this sublime science; and although at present a knowledge of the subject was confined to the few, he could not but consider that this arose in a great measure from a misconception of the acquirements which an individual should possess before venturing upon its study. . . . In conclusion, we strongly recommend all our readers who are anxious to familiarize themselves with the wonders of the heavens, to visit the Polytechnic Institution."[57] Although I use the word "rivalry" to describe the relations between Adams and Bachhoffner, no evidence shows that the two had any acquaintance with or commented on each other. The reason for suggesting they were rivals is that they both engaged in astronomy lecturing in London and enjoyed great popularity. It was unlikely that they had never heard of each other. In later years, intriguingly, Bachhoffner moved his Lenten astronomical lecture to afternoons at one or three o'clock, while Adams continued his accustomed evening performances between eight and ten.[58] Bachhoffner's scheduling change was perhaps intended to avoid a direct clash. The arrangement of different lecturing hours also suggests a deliberate distinction between the two events. A day lecture was more like a normal class in a school or scientific institution; it was particularly convenient for pupils accompanied by their teachers. In contrast, evenings were the conventional hours for theater and leisure; evening events could also draw more working-class or family audiences.

Bachhoffner left the Royal Polytechnic Institution in August 1855, when he turned his attention to another ambitious project. The Colosseum at Regent's Park, an amusement attraction that first opened in January 1829, was famed for its display of a colossal panorama of London, but it was constantly disrupted by financial crises. Its ownership changed several times and it opened for business intermittently. Bachhoffner was involved in a plan to revive the once glorious Colosseum.[59] With a good reputation as the hitherto "universally-known conductor of the Polytechnic Institution," Bachhoffner made a strong promise to the shareholders of the newly-formed company taking over the property.[60] The renovated Colosseum, under Bachhoffner's management, reopened on December 26, 1856. The *Standard* reported the event and praised the Colosseum's magnificent interior, which was undamaged after years of inactivity. The journalist indicated: "Dr. Bachhoffner proposes to add to the well-known attractions of the place lectures on various popular topics, besides a series of dissolving views, and other optical illusions."[61] Bachhoffner planned to bring his scientific repertoire from the Polytechnic to the Colosseum. To achieve this goal, he even proposed spending £2,000 to purchase scientific equipment.[62]

After the reopening of the Colosseum, Bachhoffner presented his astronomical lecture at the new site in the following Lenten season. He improved the ingredients and added new seasoning. The advertisement announced: "On the evenings of Wednesday and Friday, during Lent, Dr. Bachhoffner will deliver a lecture on Astronomy, illustrated by a new and beautiful Orrery, with selections from Haydn's oratorio of the 'Creation,' by the Crystal Palace Orchestra."[63] At the Royal Polytechnic Institution, the vacancy of astronomical lecturing was later filled by Bachhoffner's former colleague John Henry Pepper, who quickly proved to be a genius and celebrity of scientific entertainment.[64]

Bachhoffner's efforts to revive the Colosseum, however, failed both commercially and scientifically. His strategy of reducing the admission fee to 1 shilling for an all-inclusive deal was, according to the manager's claim, initially effective. It provided an affordable price for working-class visitors and attracted upward of twelve thousand visitors in a week. A *Morning Chronicle* journalist observed that scarcely half this number of visitors had passed through in a year during which the old admission charge of 4 shillings and sixpence was still in place.[65] Nevertheless, it is questionable whether these large audiences lasted long. Just one year after reopening, the company was dissolved and Bachhoffner became the sole lessee and manager.[66] He eventually sold the Colosseum in 1862. The new license holder and manager changed the program, adding more musical performances and shadow pantomimes.[67] The scientific character of the renovated Colosseum during Bachhoffner's term of office had faded away. All these changes, however, could not save the Colosseum from downfall. It soon closed, never to reopen, and the buildings were demolished in 1875.

After his unsuccessful stint at the Colosseum, Bachhoffner seems to have retired from lecturing, too. No known record indicates any lecturing activity of Bachhoffner afterward. Until his death, he retained the post of superintendent registrar in Marylebone district in London, which he had served since 1853.[68] His scientific ingenuity remained active: he was involved in an inquest into an underground railway accident in 1867, and even obtained a patent on an improved gas-fuel lamp in 1871.[69] However, the former manager of the Royal Polytechnic Institution and the Colosseum had faded from the memories of audiences. Like the Colosseum, the once glorious Royal Polytechnic also declined, finally closing for business in 1881.

The Royal Polytechnic Institution and the Colosseum shared a common fate among numerous science exhibitions in the Victorian era—proprietors and managers were constantly compelled to deal with financial difficulties. These establishments also needed to adjust their operations to adapt to the elusive and changing tastes of audiences. Many of these venues initially had an ambitious objective to integrate arts, inventions, and technology through their connections with science and to introduce these useful novelties to the public. However, as an insightful remark from a bestselling guidebook *Mogg's New Picture of London and Visitor's Guide to It* [sic] *Sights* (1844) indicated when introducing the Adelaide Gallery, "Science alone . . . had not of itself attraction sufficient for the multitude, and a resort to meretricious aid was the consequence."[70] Managers usually had to add more entertainment to attract audiences, thereby compromising and sometimes even losing the original scientific purpose of the exhibition.

Affiliation or Independence

The distinct career trajectories of Bachhoffner and Adams suggest how culturally diverse Victorian popular astronomy lecturers were. Their vocational backgrounds illustrate their dissimilarities: Adams was a schoolmaster while Bachhoffner was a chemistry lecturer with a doctoral degree. The difference was also apparent in the venues where their activities took place: the auditoriums of the West End theaters, the Royal Polytechnic Institution, and the Colosseum shared similar as well as distinctive characters. Though the two lecturers were both recognized by the public as legitimate interpreters of astronomy, they seemingly lacked any connection with astronomical establishments. These facts raise the question of the identity of astronomical lecturers: What status did they occupy within the scientific community in early Victorian Britain?

Before we go any further, it is necessary to understand the milieu of nineteenth-century British astronomy. Unlike the Continental European mode of a state-conducted, centralized governing body employing university-trained professionals, in Britain astronomy had maintained a strong "Grand Amateurs" convention.[71] A profession in the early Victorian sense, according to Jack Morrell, was a "vocation in which a professed knowledge of some aspect of science or learning was applied to human affairs or in the practice of an art founded upon such knowledge." This definition emphasizes the aspect of vocational pursuit. In other words, a professional is a person who can earn a living from a career that requires esoteric knowledge and a high level of education or training. By this definition, the early Victorians recognized few professions—merely divinity, law, medicine, and the military.[72] Science was hardly a profession during the first half of the nineteenth century. At the time, a practitioner of science preferred to be denoted as a "man of science," a phrase implying a gentleman of independent means, who was devoted to fulltime pursuits of science but not dependent on them for a living. As James Secord observes, gentlemanly science was "an ideal as much as a reality."[73] To be a gentleman meant having the freedom not only to engage in a chosen career but also to comply with a set of respectable ethics and conduct. The prevalent culture of gentlemanly science was based on the ideals of gentility in a hierarchical society to ensure the independence, virtue, and credibility of scientific practices.[74]

In early Victorian Britain, astronomy, like other branches of science, remained the playground of gentlemen. Allan Chapman indicates that adequately paid astronomical posts were few and far between in this period. He uses the term "Grand Amateurs" to describe the elites in the astronomical community.[75] Grand Amateurs were wealthy, private, independent practitioners, who were devoted to astronomical observations and research and used their own telescopes, instruments, and observatories. To support their astronomical pursuits, some had another profession or business; others amassed a fortune through inheritance or marriage. For instance, incomes from their original jobs enabled Francis Baily (a stockbroker), James South (a surgeon), William Lassell (a brewer), George Bishop (a vintner), William Henry Smyth (a naval officer), and James Nasmyth (a mechanical engineer) to indulge their interest in astronomy. John Herschel, a science baron and the foremost gentleman of science in his time, benefited from his family's fame and fortune; Lord Rosse (William Parsons, 3rd Earl of Rosse) was a member of Parliament and inherited a peerage before he built the famous six-foot telescope.[76] These Grand Amateurs possessed private resources allowing them to work on their studies while their livelihood did not depend on stargazing. The Royal Astronomical Society, the main establishment where Grand Amateurs met, read, and shared information, was a clearinghouse for independent enthusiasts rather than a powerfully centralized institution controlling and conducting research.

One rare exception among the astronomical elite is George Biddell Airy (1801–92), who lived solely on the income from his appointment as Astronomer Royal. Because of Airy's extensive involvement in public services, governmental businesses, and the renovation of Greenwich Royal Observatory's infrastructure, historians describe him as a "scientific civil servant."[77] The longevity of Airy's forty-six-year tenure as the Astronomer Royal (1835–81) made him a leader in the British astronomical community and a key science adviser to the government. He had pervasive influence and authority, whether at the level of the conduct of astronomical research or the construction of a social network facilitating careers within the community.[78] Other directors of public or university observatories usually had additional clerical or academic duties that provided their main income. Roger Hutchins summarizes the posts and salaries of astronomers in private, public, and university observatories in the nineteenth century. However, due to the burdens of teaching in universities and the shortage of grants, most of the academic astronomers—directors or professors at university observatories—had less power and influence over British astronomical communities than Grand Amateurs and the

"Greenwich-Cambridge Axis" network, which was centered around Airy and his allies at the University of Cambridge.[79]

Salaried assistants hired by individuals or observatories were another important part of the workforce in British astronomy. Observatories, whether private or public, needed assistants for maintaining day-to-day operations. The Greenwich Royal Observatory was the most prestigious employer in this job market. Aside from six senior warrant assistants, Greenwich employed and trained a few middle-class lads as "Supernumerary Computers" to do routine calculation or observation work. A journalist described this supernumerary system as a "band of scientific clerks."[80] Nonetheless, only a few assistantship positions were available in the entire country, and these professional assistants also had limited career prospects. According to Hutchins, the number of professional astronomers across the British Empire in 1860, including posts in overseas observatories such as the Cape of Good Hope, Sydney, and Melbourne, totaled only forty-two. Even including those "Computers," there were just sixty to seventy salaried positions.[81] This was the case in 1860; the community of astronomical professionals was even smaller at the beginning of the nineteenth century. These professionals were usually "invisible" in the forum of Grand Amateur science and very few of them could rise to a socially recognized managerial rank.[82]

Adams and Bachhoffner had seldom or never been involved in the astronomical community described above. They were neither Grand Amateurs nor professional employees working for observatories. None of them had affiliation with astronomical establishments such as the Royal Astronomical Society. Adams might have engaged in some astronomical observation in his early career or spare time, but there are very few accounts that he did so.[83] Bachhoffner, whose activities were closer to the sense of "man of science," was affiliated with the discipline of chemistry or experimental electrical science rather than astronomy—he was a fellow of the Chemical Society of London (from which he often styled "FCS" in the advertisements) and a member of the short-lived London Electricity Society.[84] Both men's involvement in astronomical science focused on public lecturing rather than conducting original research or routine observation. This distinguishes them from most practicing astronomers. Grand Amateurs and observatory professionals were not dependent on public lecturing for

FIGURE 3.2. Timeline of popular astronomical lecturers in London and vicinity between 1800 and 1870. This diagram shows a lecturer's approximate active period. The start and end are according to the earliest and latest record of lecturing or the date of death.

supporting their livelihoods and scientific pursuits. The astronomical elites would sometimes present addresses or lectures to institutions, as Herschel did at meetings of the BAAS, or as Airy did at the Royal Institution, but these occasions were not on a regular or commercial basis.[85] In contrast, lecturing was a business or a vocational pursuit for people like Adams and Bachhoffner. They made a profit from delivering public lectures.

Adams and Bachhoffner were just two of the lecturers active on the metropolitan lecturing scene. Numerous lecturers were active in London's vibrant science marketplace during the first half of the nineteenth century. Figure 3.2 is a timeline of several astronomical lecturers in London and its vicinity between 1800 and 1870. It shows each lecturer's approximate active period according to the earliest and latest known lecturing dates or the year of the lecturer's death, as derived from advertisements, newspapers, and biographical accounts. These lecturers regularly delivered public discourses on astronomy. Many of them had also competed in the seasonal market of astronomical extravaganzas during Lent. Some, like Adams, were active for a long period and enjoyed consecutive years of lecturing. Some did not stay in the business long, yet they still had a significant impact on the market. For example, George Bartley's lectures at the English Opera House received critical acclaim and were frequently mentioned by contemporaries, though his astronomical sideline was only staged for less than a decade. Some associated with scientific institutions, like Bachhoffner's connection with the Royal Polytechnic. John Wallis was another example, as he was employed by various organizations including the Royal and the London Institutions.

Figure 3.2 also shows a rough trend of the astronomical lecturing trade. There was a period of growth during the 1820s and 1830s, as new entrepreneurs continued entering lecturing ventures. However, following the retirement or death of some prominent names that enjoyed popularity in the early Victorian era, the trade drastically declined in and after the 1860s, especially for those private entrepreneurs like Adams. Notwithstanding this eventual decline, I argue that public astronomy lecturing was a shared arena of private and institutional lecturers by the mid-nineteenth century.

My argument responds to two pioneering studies of the metropolitan lecturing milieu and popular astronomy by J. N. Hays and Ian Inkster, respectively. Hays claims that scientific lecturing in London was decisively institutionalized by the 1820s, when metropolitan lecturing scenes changed and became formalized. Institutionalization eclipsed private lecturers and propelled the professionalization of the men of science.[86] Hays's observation is based on the establishment of scientific institutions in the first two decades of the century and the increase of lecturers' affiliation with those formalized bodies. Similarly, Inkster indicates an institutional transition in popular astronomy in the early nineteenth century, when academic men ousted old-school independent lecturers from the institutions.[87] However, these two observations do not give the whole picture. The growth of institutionalized science did not immediately overwhelm private lecturers in the market. The private astronomy lecturing trade continued to thrive until the 1860s.

Lightman warns that the demarcation between popularizers and practitioners should not be seen as too rigid. The dichotomy between institutional and private lecturers was also porous. The first problem is the supposedly clear distinction between scientific practice sites and performance venues. The former, such as learned societies and the Royal Institution, produced knowledge through conducting scientific research, experiment, and innovation; the latter, including science exhibitions and museums, purveyed knowledge to the broader public by entertaining displays.[88] However, institutions with a hybrid mix of innovative experiments and spectacular performance, like the Royal Polytechnic Institution, where sparks from electrical machines and the illusory extravaganza of "Pepper's ghost" were demonstrated, challenge our received demarcations. This dilemma over how to categorize a site as a scientific space parallels Lightman's emphasis on the difficulties of clearly distinguishing popularizers from practitioners. Another problem is the extent to which a lecturer was affiliated with an institution. Lecturers might give presentations elsewhere despite being formally associated with a particular institution. These freelance institutional lecturers managed their businesses the way private entrepreneurs did. They hovered in gray areas of varied status. Moreover, lecturers might change their affiliation during a career. Bachhoffner, for instance, had firm institutional connections with the Polytechnic before switching to the Colosseum as an independent entrepreneur.

These potential problems reflect not only the practical difficulties of categorizing individuals and sites but also the changing character of science throughout the nineteenth century. What later became known as professional and popular science in the twentieth century and thereafter, as Lightman points out, did not exist in the Victorian period. We should also be cautious about Chapman's use of the term "Grand Amateur." Despite providing an impressively vivid portrait of nineteenth-century astronomers, the designation bears anachronistic connotations of contradistinction to the modern meanings of either the positive qualities associated with "professional" or the negative associations with "amateur." The terms "professional" and "amateur" did not indicate a distinction between insiders and outsiders of science for nineteenth-century contemporaries, nor did superiority in expertise or place in the hierarchy.[89] Ambiguity concerning professional and popular science also relates to the spaces in which science is conducted. Astronomy lecturing, too, was carried out in various sites, as we have further explored in chapter 2.

Rather than misconceiving the professionalization of science as an inevitable, uniform, and universal process, historians now consider it a contingent process in which scientific practices, boundary marking, and social relations were changing in an industrializing society.[90] Scholars in recent decades have also indicated that science in the nineteenth century was a contested space, whether in the geographical or cultural sense, where groups and individuals shaped the identity of scientific practitioners and competed for the legitimacy of what counted as science in society.[91] The cases of C. H. Adams, Bachhoffner, and many other astronomical lecturers demonstrate that the conventional separation between professional and amateur practitioners is inapplicable when analyzing the context of astronomy popularization in the nineteenth century. Alternatively, the categories of institutional and private lecturers, despite their porosity and the risk of oversimplification, are useful in assessing the activities of popularizers.

Prospects for a Career and Prosperity

What motivated numerous people to enter the astronomy lecturing business? Studies of nineteenth-century popular science publishing pay great attention to the political or religious agenda behind the popularizers. The cases that scholars have examined include the publications of the Society for the Diffusion of Useful Knowledge, which coincided with the Whig ideal of social reform.[92] In addition, the Bridgewater Treatises explored natural theology, as did the works conducted by the evangelical Religious Tract Society.[93] Astronomy was an unrivaled subject for evoking the imagery and sentiment of the sublime; hence, religious rhetoric was common in popular astronomy writings and lecturing. In the meantime, the order of the motions of heavenly bodies and the materialistic nebular hypothesis could each lead to completely different views on current sovereign and religious authorities. Nevertheless, it did not necessarily mean that a lecturer had a particular religious or political agenda. The prospects for making profits and promoting status were two important motivations for engaging in astronomy lecturing.

Secord coined the term "commercial science" to designate a wide range of scientific activities involving trade and profits in the marketplace. Science was, as Secord asserts, part of a commercial culture of exhibition. Opportunities for making money from the display of knowledge had been drastically transformed and increased by the revolution in communication in the industrial era.[94] These opportunities covered a variety of jobs that we might associate with popular science today, such as authoring, editing, lecturing, showmanship, specimen dealing, instrument making, museum curating, and so on. Secord argues that the term "commercial science" is more useful than "professional science" or "popular science," which easily lead to modern positive or dismissive connotations, respectively.[95]

In addition to making profits, commercial science provided a potential career path to science in the preprofessional period. As I have discussed in the preceding section, science in the first half of the nineteenth century remained a gentlemanly pursuit. People who wanted to cultivate a career interest in science required independent means. For a practitioner without independent means, engaging in commercial science could enhance visibility while simultaneously earning a living, and, moreover, it could offer a chance to join the genteel circle and clubland of science. Michael Faraday, who started his scientific career as a laboratory assistant, was a classic example of someone rising from the humblest circumstances to a preeminent position at the Royal Institution. Private lecturers who operated extramural hands-on courses in practical chemistry and botany for medical students exemplify another manner of exploiting commercial science.[96] For those with a more genteel pedigree, engaging in commercial science was a means to maintain their scientific research when financial pressure loomed, but they were pleased to avoid the drudgery of employment—even a scientific job—if they could. For example, the botanists William Hooker at the University of Glasgow and his friend John Lindley at University College London had to supplement their meager salaries by writing for popular horticultural magazines.[97] Edward Forbes at King's College London, too, was troubled by the same financial problems, which led to his extra service in "mill-horse drudgery" at the Geological Society and any other "hackwork" that he could obtain.[98]

Public lecturing was not an unusual route for a learned young man wanting to establish a scientific reputation in early nineteenth-century Britain. It was also deemed a steppingstone to proving their mettle. A wide spectrum of institutions, ranging from the high-end Royal Institution to the mechanics' institutes, aimed at the working classes. Permanent posts were scarce, but temporary lecturers for delivering courses were in demand. Charles Babbage, for example, lectured to the Royal Institution on astronomy in 1815. This was his first employment immediately after graduating from Cambridge.[99] Babbage then married and the couple later moved to London. As a junior scholar who wanted to get a foothold in the metropolitan scientific circle, this lecturing employment at the Royal Institution was a good test of his comprehension of the subject. Another example, though this was more of an honorary guest lecture rather than employment, was the lecture titled "Astronomy: The Scale of the Solar System," by John William Strutt in 1865 at Witham Literary Institution. Strutt, who later succeeded to a peerage as the 3rd Baron Rayleigh and became a Nobel laureate in physics, was then a young Cambridge student who had just obtained the honor of Senior Wrangler, the highest scoring student in the Mathematical Tripos competition. In recognition of Strutt's achievement and peerage, the president of the institution personally chaired the lecture. He praised Strutt, "our conquering hero . . . the whole county of Essex rejoices in Mr. Strutt's success." According to the journalist, Strutt's lecture attracted "a very numerous and highly influential company."[100]

To get access to scientific coteries and maintain a scientific career might not be a prime objective for artisan lecturers, but they were certainly interested in improving their social status through commercial science. Successful lecturers on astronomy from humble working-class backgrounds exemplify the potential for prosperity in the lecturing business. One representative figure was John Bird, whose lecturing career started after 1814.[101] Before 1814, probably when in his early twenties, Bird was a journeyman carpenter in Abingdon, Berkshire. Despite lacking a formal education, Bird made astronomical instruments such as a tellurian—a device that demonstrated the tilted rotation of the Earth—by using "an old print on a leaf of Ferguson's Astronomy."[102] After the encouragement of an unnamed patron and a successful exhibition of his instruments at the town hall, Bird abandoned his carpenter's trade and found a vocation in lecturing.

The story of a genius with humble origins interested journalists, as two magazines reported: "An extraordinary instance of innate scientific genius has been lately evinced in the person . . . Bird, who, less than a twelvemonth since, followed the humble occupation of a journeyman carpenter. . . . He has since delivered lectures, with astonishing perspicuity, in the principal towns of Berkshire, Wiltshire, and Somerset-

FIGURE 3.3. Handbill of John Bird's lectures at Charter House, 1830. Science Museum, inventory no. 1980-930/4. © The Board of Trustees of the Science Museum, London.

shire."[103] By the 1820s, John Bird was already a well-established private lecturer. He was employed to lecture in public schools including the Charter House, Westminster, and Eton (fig. 3.3).[104] Judging by the relatively high admission charge (3 shillings for each lecture, which was equal to the price of a box seat in a West End theater) to Bird's courses at the Charter House, he must have had a wealthy audience. Bird enjoyed fame in aristocratic circles as the astronomical preceptor of the Marquis of Douro, the eldest son of the Duke of Wellington, and he received the patronage of King William IV. A posthumous biography in the *Leisure Hour* describes Bird thus:

> He certainly had little learning; his qualifications consisting in reverent admiration for, and enthusiastic ardour in pursuing and illustrating, astronomical truths. Moreover, he possessed an inventive mind, a retentive memory, genuine natural humour, versatility, and readiness. There was, however, a want of refinement in his speech and manner. Still, notwithstanding these drawbacks, in those days his capacities were sufficient to insure him the reputation of a public favourite. His lectures were always extemporaneous.... Mr. Bird, in truth, raised a host of imitators, though none of them possessed the originality of his mind.[105]

In terms of patronage, Bird followed the traditional pattern of instrument makers and astronomers who were autodidacts in the eighteenth century, such as George Adams, James Ferguson, William Herschel, among others. They were all of obscure origins, established good reputations by virtues of their crafts and expertise, and eventually acquired royal patronage.[106] Bird's ingenuity and self-improvement, regardless of his humble background, also made him a good example for Victorian readers. The title of the *Leisure Hour* article appropriately indicates that Bird was a "lecturer of the old school." This assessment suggests a traditional value represented not only by his lecturing style but also by his aristocratic patronage and endorsement.

Bird's success directly inspired another astronomical lecturer, Robert Children, who was originally a small-scale master bootmaker in Bethnal Green, London. After attending a lecture delivered by Bird sometime around 1835, Children began to develop his own lecturing business.[107] Like Bird, at the beginning of his career Children had to overcome many problems, including illiteracy, bad pronunciation, and ignorance of specialized knowledge. Nevertheless, he persevered and survived in the market. The *Hampshire Advertiser* cited a paragraph from the *Oxford University Herald* of November 28, 1835, to report Children's lectures in the town hall of Oxford, which were "well attended, particularly those in the morning given at the request of several of the heads of Colleges."[108] The *Essex Standard* also praised Children's lecture in Sudbury in 1841, which was under the patronage of the mayor and an Anglican minister. The journalist made the criticism that most scientific lecturers failed to impart knowledge to the audience, but this was "not the case with Mr. Children.... Without any pretensions to the higher flights of oratory, Mr. Children, in easy and familiar language, explained the wonders of heavenly bodies, and illustrated his subject with such a variety of transparencies, and an ingeniously-contrived mechanical apparatus, constructed by himself, on a new and improved principle, that his most difficult theories must have been understood even by a mind not previously turned to the consideration of astronomy."[109] Children's success in the lecturing business made him quit his original boot-making trade. He eventually acquired a fortune of £6,000 from astronomy lecturing and retired as a "great farmer" in America.[110]

The cases of John Bird and Robert Children exemplify not only the potential prosperity of the private lecturing trade but also the value of self-improvement. "Improvement" was a widespread idea in Georgian and Victorian Britain. It had many levels of meanings on different scales, ranging from a land to an individual. In a more grandiose sense, improvement could mean the morality of a government taking command of natural resources effectively, to the economic benefit of the nation and the empire. This was often seen in contemporary rhetoric that supported state initiatives whether at home or involved in imperial expansion.[111] For individuals, self-improvement was a recurring theme frequently addressed by Victorian writers. An individual could improve knowledge, morals, and faith through industrious self-learning; through self-improvement, one could take advantage of personal talents and thus make a better, more useful life.[112] Cultivation of science was deemed to be a good use of the divine gift of reason, and was a "dignified means of excluding those modes of abusing time which are the sin and disgrace of many young persons."[113]

In addition, self-improvement encompasses the virtues of hard work and perseverance, along with the prospect of consequent rewards. The success of Bird and Children is therefore similar to the famous story of Faraday, who transformed himself from a humble bookbinder's apprentice into a recognized man of science. These cases all reflect the climb from a lower social status to a better vocation with erudition, respectability, and material prosperity. It is not surprising that many Victorian accounts, such as the *Leisure Hour* article about Bird, were full of similar uplifting stories. Like the moral messages conveyed in their astronomical lectures, the lecturers' own lives were themselves a good example to the audience.

When William Walker began taking over his father Adam Walker's lecturing business in the last two decades of the eighteenth century, the scientific communities in Britain had not yet been institutionalized on a massive scale. At the time, the Royal Society was one of the few establishments dominating the landscape of scientific institutions, although some metropolitan and provincial learned societies had also started burgeoning. Meanwhile, public lecturing of science remained a largely private business, as lecturers like Adam and William Walker did not necessarily affiliate with scientific institutions and they usually ran their lectures independently. Moreover, many astronomical lecturers were self-educated artisans who entered the trade because of their mechanical genius, or because they sought material rewards or higher social status. Some, such as John Bird, got recognition and patronage from aristocrats. Their successful stories also became inspirational models for self-help and moral education. Lecturers of this type still existed in the early Victorian era, and we can consider them a continuation of the public philosophical lecturing trade typical of the eighteenth century.

However, with the rapid growth of scientific institutions in the first two decades of the nineteenth century, more and more lecturers were employed by or associated their lecturing activities with scientific establishments after the 1820s. The examples of John Wallis and George Bachhoffner illustrate this institutionalized trend in popular astronomy. The drastic and complete institutionalization of scientific lecturing around the 1820s as claimed by J. N. Hays nevertheless did not occur. The activities of private entrepreneurs such as C. H. Adams after the 1820s showed that there was still ample room for independent astronomy lecturing. The distinction between private and institutional lecturers only became apparent by the mid-nineteenth century, but the lines remained blurred and the two were sometimes interchangeable. Popularizers of astronomy, with or without institutional affiliations, established their scientific prowess by building positive audience reception rather than securing academic qualifications in advance. In any case, very few commercial astronomical lecturers in the mid-nineteenth century were practicing astronomers who conducted original research and made regular observations. They were neither Grand Amateurs nor observatory professionals and they were hardly part of the system of gentlemanly science.

CHAPTER 4

SUBJECTS

Imagine being a lecturer designing a syllabus or a program on astronomy for your debut. You are expecting a curious audience, most of whom know little about the sky. Some are looking forward to enhancing their understanding of the universe intellectually or morally, while others might just want to find some entertainment during a humdrum day. What kind of topics would you put into your presentation? A satirical article titled "Hints to Lecturers" in 1808 offered a humorous answer. Under the pseudonym "Crop the Conjuror," the author tartly reviewed many different disciplines from botany to geology and gave "advice" to those gentlemen in the trade of philosophy. Astronomy was also on Crop's list. "If astronomy is your forte, I must acknowledge that you have an extensive field," Crop explained, "[you] may tell us all that the stars have been doing these thousands years, for nobody can contradict you; then after leading us to the *farthest verge* of *boundless* æther, you may descend to a few private anecdotes of the solar system, talk of the conjunctions of Venus and Mercury, and tell us, with a smile, that, with the exception of the *conjunctions* and *oppositions*, there are no other signs of matrimony in the heaven, but Saturn's *ring* and Lunar's *horns*!"[1] Astronomical lecturing in reality, of course, was not as absurd as Crop the Conjuror suggested. Some common topics were similar to those in today's middle-school science classrooms, such as the motions of planets, the causes of eclipses and seasons, and the nature of mysterious comets. Some discourses, however, were quite different from those in present-day curriculums, which reflected issues that concerned the public in the nineteenth century.

SUBJECTS

This chapter discusses the subjects regularly appearing in the syllabi of nineteenth-century popular astronomy lectures. They were often a mixture of convention and novelty. Conventional topics that had frequently been delivered since the eighteenth century, such as the physical features of the Earth and the Sun, the structure and motions of the planetary system, and the causes of eclipses, tides, and seasons, continued to be fundamental materials in the syllabi. On the other hand, some of the latest discoveries, such as new planets, made perfectly sensational headlines in advertisements or playbills. There was also room for some controversial subjects debated by theologians and men of science, including the plurality of worlds and life on another planet. These subjects could contain radical and potentially subversive messages.

The most direct sources for revealing the topics covered in nineteenth-century popular astronomy lectures, of course, are the syllabi of lectures. Like their predecessors in the eighteenth century, some private lecturers were especially keen to publish treatises or pamphlets on astronomical subjects for the purpose of promoting their businesses, which often contained the outline of their lectures. D. F. Walker, for example, continued to publish his eldest brother William's annual syllabus for the eidouranion lecture, which reached its thirty-first edition in 1824. Theaters or astronomical lecturers who used large transparent orreries also issued posters or handbills for advertisements, which often listed programs or scenes that were going to be performed and functioned like a syllabus. Institutional lecturers usually did not need to worry about publishing their lectures because they were often covered, sometimes in detail, by contemporary newspapers or periodicals. The Royal Institution's own series of journals—their titles changed several times, and they were published intermittently—regularly included a synopsis of the institution's courses of lectures or Friday Evening Discourse events. Other magazines, such as the *Athenaeum* and the *Literary Gazette*, would also report on what was covered in the Royal Institution's lectures.[2] Furthermore, scientific institutions also frequently printed and sent the latest syllabi to their members and subscribers. In addition to lecture syllabi, some illustrative works of popular astronomy or related lantern slides, though not necessarily linked to specific lectures, are indications suggesting popular topics in astronomy lectures. The above materials are all sources for this chapter's discussions on subjects of popular astronomy lectures.

SUBJECTS

Syllabus in Common

The juvenile lectures at the Royal Institution, later branded as the Christmas Lectures, have been the institution's trademark attraction since 1825.[3] Each year, the Royal Institution invites a scientist to present a course of several lectures around the two weeks during the Christmas and New Year's holidays. Lecturers who stood on the podium for the series have included many big names such as Michael Faraday, John Tyndall, Carl Sagan, and David Attenborough. When John Wallis was invited to give the juvenile lectures in 1846, he faced a task that was familiar to him but not simple. The task was familiar because it was not Wallis's first time—he had already given the juvenile lectures at the Royal Institution in 1826 and 1838, respectively. For the third time, Wallis titled his lecture "The Rudiments of Astronomy," promising that the course would contain the simplest, most basic facts about the sublime science. This was not an easy undertaking because he had to organize a course of six, two-hour lectures for youngsters. It would be quite a challenge to cram the whole universe, despite the rudiments, into a mere six lectures—not to mention that the juvenile audience was easily distracted and might fail to comprehend abstruse ideas.

Wallis nevertheless did a decent job. A copy of the syllabus of Wallis's juvenile lectures is preserved in the Royal Institution's archives. The syllabus was printed on paper cards, which were usually sent to the institution's members and subscribers as an advertisement.[4] Each lecture's subjects were listed with brief descriptions in the syllabus. The first lecture started with the general phenomena of the Moon, including the lunar orbit and movement, and then connected to the causes of eclipses. Wallis used a "Picture of the curious facts observed during the total eclipse of the sun, as seen in Italy in 1842" as the concluding topic. This curious picture was probably related to the solar corona and prominences, or the phenomenon of the "Baily's beads" effect, created by the irregular topography of the lunar limb and first explained by the English astronomer Francis Baily. Baily observed this total eclipse at Pavia in northern Italy, and his expedition had attracted public attention to the eclipse.[5]

The second lecture focused on the Earth, including its figure, rotation, and annual revolution. It concluded with the phenomenon and cause of the seasons. "Time" was the theme of the third lecture with many relevant topics, including the computation of time; the use of sundials; the Meridian line; differences between sidereal and tropical years; and leap years and the reformation of the calendar. It was notable that Wallis arranged "time" as the theme on this occasion, since the lecture was to be delivered on New Year's Eve. The last day of the year would be a perfect occasion to discuss the natural and artificial divisions of time and calendar. The fourth lecture focused on the doctrine of universal gravitation, including the nature of planetary motions, the laws of falling bodies, and the precession of the equinox. The fifth lecture concentrated on the tides and the relative forces of the sun and moon as causes of the phenomena.

Finally, in the sixth lecture, Wallis took the audience away from the Earth to outer space, as he conducted a general survey of the solar system including all the planets, comets, and telescopic appearances of these celestial objects. He also introduced the "newly-discovered planets," which certainly included the hottest headline Neptune, which had just been discovered in September 1846. Planetoids such as Ceres and Vesta, both discovered within the century, were likely also included. The concluding topic eventually brought the audience back to their home. Wallis introduced the revolving planisphere of stars, which is visible in London, a practical ending that encouraged his juvenile audiences to explore the night sky themselves.

The syllabus of Wallis's juvenile lectures is an excellent example of astronomical lecture subjects in early Victorian Britain. This syllabus contains many subjects in common with other contemporary astronomical lectures. The phenomena related to the movements of the Earth or the Moon, such as eclipses, seasons, and tides, were recurrent topics in conventional astronomical lectures during the first half of the nineteenth century. For instance, the lectures of both D. F. Walker and C. H. Adams encompassed these subjects in their repertoire.

A typical Adams lecture in a theater was often divided into three parts: the Earth, the Moon, and a survey of cosmological models from the Ptolemaic to the Copernican systems.[6] The first part concerning the Earth included its shape, dimensions, motions, and the cause of the seasons. The second part focused on the phases, orbit, and motions of the Moon, with a special feature on the phenomena of tides, which was explained using a "Mechanical Transparent Apparatus, of Novel

Construction." The third part covered the whole solar system, "The Science considered systematically, the Ptolemaic, Egyptian, Tychonic, and Copernican—the Copernican the only true one—Telescopic Views of the Planets—comparative magnitudes, &c." Adams maintained this three-part organizational structure for his lecture throughout his career.[7] Nevertheless, since Adams had a long career spanning more than thirty years, some updates were necessary from time to time, such as adding newly discovered planets into his orrery.

In comparison with Adams's lecture syllabi, D. F. Walker's were further fragmented and contained more detailed descriptions. Walker usually divided a lecture into five parts—he called each part a "scene" and thus fashioned a more theatrical look for his lecture.[8] A typical D. F. Walker's lecture would consist of the following scenes:

Scene 1.—Exhibits the EARTH in ANNUAL and DIURNAL Motion: Day, Night, Twilight, long and short Days, the Seasons, Years, &c. &c. are rendered so plain and intelligible, that a bare inspection of the Machine explains their cause, to any capacity. These Phenomena are explained on a new Transparent Globe, two feet diameter, revolving on its inclined axis before the Sun and through the Zodiac, to produce the Seasons. The Stars composing the figures surrounding the whole.

Scene 2.—Exhibits the EARTH and MOON.—In which the cause of her different Phases, or change of appearance; her Eclipses and those of the Sun; with the view of her Disk, as seen by the most powerful Telescopes, are the principal Features.

Scene 3.—The TIDES.—Exceptions reconciled, &c. particularly those in the Irish Sea.

Scene 4.—Has every PLANET and SATELLITE in ANNUAL and DIURNAL MOTION at once; a COMET descends from one Side, and retires at the other of the Machine, having its motion accelerated or retarded according to the Law of Planetary Motion.

Scene 5.—The PROBABLE CONSTRUCTION of the UNIVERSE, Exhibiting every Star as a Sun, like ours of the Solar System.[9]

From this syllabus, we can easily recognize a theatrical rendition of astronomy in Walker's lecture. Walker not only called each part a "scene" but also described to the audience what scenarios they were expected to watch in detail, such as the emergence of a comet on the stage in Scene 4. Some of Walker's early playbills even noted which piece of

TABLE 4.1. SUMMARY OF CONTENT FROM THE SETS OF ASTRONOMICAL LANTERN SLIDES, ILLUSTRATIVE BOOKS, AND LECTURE SYLLABI

Subjects	Astronomical lantern slides*		Illustrative books**	Lectures			
	W. & S. Jones (ca. 1820s)	Carpenter & Westley (ca. 1830s)	Blunt, *The Beauty of the Heavens* (1842)	D. F. Walker (1824)	George Bartley (1827)	C. H. Adams (1839)	John Wallis (1846)
Rotundity of the Earth	•	•	•	•	•	•	•
Cause of the seasons	•	•	•	•	•	•	•
Rotation of the Earth			•	•	•	•	•
The Sun	•	•	•	•	•	•	•
The Moon		•	•	•	•	•	•
The Moon (Orbit and phases)	•	•	•	•	•	•	•
Tides			•	•	•	•	•
Solar or lunar eclipses	•	•	•	•	•	•	•
Ptolemaic, Copernican, and Tychonic systems	•	•				•	•
Newtonian system	•	•	•	•	•	•	•
Universal gravitation			•	•			•
Plurality of worlds				•	•		
Relative sizes of the Sun, planets, and their orbits	•	•	•	•	•	•	•
General survey of the Solar System	•	•	•	•	•	•	•
Mercury	•	•	•	•	•	•	•
Venus	•	•	•	•	•	•	•
Mars	•	•	•	•	•	•	•

Subjects	Astronomical lantern slides*		Illustrative books**	Lectures			
	W. & S. Jones (ca. 1820s)	Carpenter & Westley (ca. 1830s)	Blunt, *The Beauty of the Heavens* (1842)	D. F. Walker (1824)	George Bartley (1827)	C. H. Adams (1839)	John Wallis (1846)
Jupiter	•	•	•	•	•	•	•
Saturn	•	•	•	•	•	•	•
Uranus	•	•	•	•	•	•	•
Neptune							•
Milky Way	•	•	•				
Nebulae		•	•				
Zodiacal constellations	•	•	•	•	•	•	

Sources: *Bush, "Again with Feeling," 499, table 1; **C. F. Blunt, *Beauty of the Heavens*; SMG: 1980-930; RAS: Add MS 88; BL: Add MS 42875; and RI: MS GB 2: 47/71C&D.

music would be played during the intervals between scenes. The English composer John Wall Callcott's hymn, "These as they change Almighty Father are but the varied God," was performed between Scene 1 and Scene 2. The air "Holy, Holy, Lord" by George Frideric Handel was placed between Scene 2 and Scene 3. The duet "The Heavens Are Telling" from Joseph Haydn's oratorio *The Creation* was performed between Scene 4 and Scene 5.[10] By choosing these religious hymns and inserting them in the middle of his lectures, Walker transformed his astronomy lectures into a new kind of oratorio that was both scientific and artistic as well as religiously inspired.

Wallis's juvenile lecture at the Royal Institution or Walker's celestial oratorio in theaters notwithstanding, the syllabi of these early nineteenth-century astronomy lectures shared some common subjects. Table 4.1 summarizes content according to several representative lecture syllabi, illustrative diagrams, and sets of lantern slides in the first half of the nineteenth century. The data of lantern slides in this table are based on Martin Bush's surveys.[11] *The Beauty of the Heavens* (1840) by Charles F. Blunt exemplifies illustrative books of astronomy. Embellished with 104 full-page color illustrations, *The Beauty of the*

Heavens incorporated elegant diagrams into a concise introduction of astronomy and was hailed as a "family astronomical lecture" by its author. Contemporary popular astronomy books or sets of astronomical lantern slides were sometimes closely arranged according to an actual lecture as a synopsis or a supplement, and hence they were also invaluable references showing common subjects.[12]

The common subjects in popular astronomy lectures reflected the significant influence of eighteenth-century public philosophical lecturing, which had two significant characteristics: they were instrument-oriented and based on Newtonian science. Demonstrations of experimental philosophy relied on various instruments ranging from the air pump to thermometers. The design of a course usually centered on the use and display of philosophical apparatus. This instrument-oriented principle also applied to the subject of astronomy: apparatus such as globes, orreries, telescopes, and sundials, were all common items on a philosophical lecturer's equipment list. Just as Adam Walker passed his lecturing apparatus and businesses to his sons, nineteenth-century astronomy lecturers inherited the same convention from their eighteenth-century predecessors, but with improvements: bigger and more extravagant visual aids. Lecturers were also not shy about assuring the audience that they would show state-of-the-art scenery and machinery.

Newtonian science was the other keynote component. The laws of motion and gravitation were Newton's great achievement, and nothing could exemplify these principles better than the revolution of the planets. In William Whewell's words, astronomy is "the queen of sciences" and "the only perfect science," since "the grand law of causation by which they are all bound together has been enunciated for 150 years; and we have in this case an example of a science in that elevated state of flourishing maturity."[13] Whewell's high praise for physical astronomy here was his taking of the publication of Newton's *Principia* (about 150 years earlier than Whewell's address) as an example to indicate that the Newtonian laws had completed the ancient science of astronomy. There were so many reasons to enshrine Isaac Newton in the nation's pantheon: his scientific genius, the national pride in his scientific achievements, and even the moral and religious virtues of his cosmological view.[14] Because Newton incorporated divine presence into his laws of gravitation and cosmological framework, his disciples regarded his scientific work as a proper homage to the Deity as well as a powerful tool for refuting Cartesian materialism and Pierre-Simon Laplace's entirely mechanical universe.[15] Narratives placing Newton at the pinnacle of all scientific geniuses were quite common in eighteenth- and nineteenth-century English literature. For instance, this view was embodied in an article titled "Faith in Astronomy" in the *London Saturday Journal*, which acknowledged and introduced the achievements of astronomy. The journalist quoted the renowned writer William Godwin to ponder: "To think with what composure and confidence a succession of persons of the greatest genius have launched themselves in illimitable space. . . . The illustrious names of Copernicus, Galileo, Gassendi, Kepler, Halley, and Newton, impress us with awe." Godwin also expressed doubt as to how astronomers came to be acquainted with the measurements of heavenly bodies. He nevertheless claimed: "Is it not enough? *Newton and his compeers have said it!*"[16] This assertion reflects Newton's preeminent status in the shrine of British science.

The influence of Newton and his disciples was pervasive in the curriculum and narratives of astronomical lectures in early nineteenth-century Britain. One of the six lectures in Wallis's juvenile lectures was devoted to gravitation and Newton's laws. C. H. Adams also claimed in his advertisement that his machinery would "render the famous theorem of Sir Isaac Newton more intelligible than any hitherto exhibited."[17] The comet scene in D. F. Walker's lecture demonstrated that "its motion accelerated or retarded according to the Law of Planetary Motion." Other forms of media also indicated the same result. For example, as table 4.1 shows, the diagram of the Newtonian system was among the common subjects for contemporary astronomical lantern slides, although diagrams illustrating universal gravitation were usually lacking. This is possibly because it was not easy to express the principles of universal gravitation simply on a slide, whereas an actual lecture or a book would be easier for delivering an adequate exposition. A set of astronomical lantern slides made by W. & S. Jones even included a portrait of Isaac Newton.[18] In any case, we find the English national hero Sir Isaac everywhere.

Astronomical lecturers loved to address the topic of solar eclipses, especially when the public's attention to such astronomical spectacles soared to a new height during the mid-nineteenth century. Three total eclipses during this period (1842, 1851, and 1860) were major sources of scientific investigations and public interest because they were visible in Europe. This enthusiasm for eclipses was not limited to Britain—European audiences were also fascinated by them. Many astronomers organized expeditions to suitable observational sites. Francis Baily's expedition to Italy during the total eclipse of 1842, for example, drew public attention, and his observation of the solar corona and prominence received extensive coverage in the newspapers.[19] The French astronomer François Arago's observation during the total eclipse of 1842 also gained public interest in France. These astronomers fueled enthusiasm for chasing solar eclipses for decades to come.[20] Depictions of solar eclipses, along with the curious phenomena that only can be seen during a total eclipse, were major attractions in astronomical lectures. C. H. Adams's lecture syllabus in 1839 advertised: "A total eclipse, and the annular eclipse of the Sun, as seen at Edinburgh, will be represented."[21] This probably referred to Baily's observation of the annular eclipse of 1836 in Scotland. Later, in the Lenten season of 1858, Adams advertised that he would "illustrate, with singular effect, the CORONA during TOTAL ECLIPSES of the SUN."[22]

In addition to the astronomical spectacle itself, changes in human perceptions of solar eclipses throughout history—from ancient people's ignorance and fears to the rational understanding of its natural cause—were classic proofs of the advancement of knowledge and human reason over superstition. As George Bartley indicated in his lecture, "In the early Ages of the World, ere men had learnt to judge of effects by their Causes, a total Eclipse of the Sun or Moon was regarded with the utmost consternation. . . . But these Opinions, as distressing as they were erroneous, are . . . entirely exploded."[23] Such narratives perfectly illustrate the close alliance between astronomy and enlightenment, and why astronomy lecturers were happy to explain solar eclipses in great detail.

NEWS OF DISCOVERY

Two of the three Friday Discourses at the Royal Institution before 1860 given by George Airy, the Astronomer Royal at the time, were related to solar eclipses (see chapter 2, table 2.2). The one titled "The Total Solar Eclipse of 1851, July 28" was delivered on May 2, 1851, about three months before the eclipse, so it was a "preview" of the upcoming astronomical spectacle. This illustrates the use of scientific news in a lecture, whether it was a recent story or an upcoming event.

Airy's lecture was evidently successful. The attendance at this Friday Discourse consisted of 610 people, thus exceeding the average size of the Friday Discourse audience in the same year, which was around 565.[24] In the lecture, Airy explained the cause, period, and shadow paths of solar eclipses. He also used historical cases to compare with the upcoming one. In his conclusion, Airy quoted an American newspaper to show the great interest in this event on the other side of the Atlantic and anticipated that it would be the "strongest inducements for Americans to visit Europe in the coming summer." Meanwhile, he trusted that his fellow English people would be interested in the total solar eclipse, too. To encourage the audience to join the eclipse chase, Airy assured potential visitors that no special skill in astronomical observation was required. Furthermore, he generously provided his professional support to the public, as "a series of suggestions for the observation, accompanied by a map" that had been prepared by a committee of which he was a member. This guide was "nearly ready to leave the printer's hands." Airy promised to pass a copy of the guide to any person who submitted an application to him.[25] Such a handbook was absolutely the hottest official guide for enthusiastic sun-chasing tourists.

Reports of recent astronomical occurrences or the latest scientific discoveries were common topics in the Friday Discourses at the Royal Institution. For instance, Charles Piazzi Smyth's lecture on March 5, 1858, "Account of the Astronomical Experiment on the Peak of Teneriffe in 1856," was Smyth's report of the world's first experiment to examine the effect of elevation on astronomical observation.[26] This expedition was supported by the British government and duly overseen by the Astronomer Royal to test how much telescopic performance could be improved in a thinner atmosphere by "elevating instruments and observers some 10,000 feet above the sea level."[27] Smyth led this expedition to the Spanish island during the summer of 1856, and the results of the investigation were transmitted to the Royal Society by the Admiralty for publication in the following year. In his lecture, Smyth did not bother the audience with numerical and technical particulars. Instead, he narrated the journey like a hunter showing his safari journal and trophies: from the organization of the expedition to the landscapes of Tenerife, exhibiting photographs made during the expedition. Smyth concluded his lecture with an idealistic remark on the "social bearing" of this astronomical experiment on the peak of Tenerife:

> The claims of sciences to respect amongst men, for its services in promoting the union of nations and the brotherhood of mankind, have been often dwelt on. Of this admirable and humanizing tendency, is not our experiment on Teneriffe an example, within its little range? See an observer sent out by the English Government, received in a fortified town of the Spaniards, not only without distrust, but as frankly as if one of themselves. And did they suffer by it? We took no notes of their forts and guns, and military array, we applied ourselves to our scientific business alone; and if we have brought away anything more from Teneriffe than what I have already had honour of describing to you, it is, respect and admiration for the Spanish character; and grand ideas of the results to astronomy as well as some other sciences, if this first experiment, this mere trial of a new method, be annually repeated, and energetically followed out.[28]

This passionate conclusion elevated the lecture to an uplifting speech rather than merely a scientific work report between peers. Smyth knew that many of his audience members were lay enthusiasts rather than scientific experts. Instead of imparting new knowledge, it was more important to arouse the public's interest in and support for science. By neglecting lots of technical details and focusing on the process and morals of the expedition, he effectively connected astronomical studies with a higher universal purpose and virtue.

Michael Faraday, too, delivered a Friday Discourse on an astronomical expedition at the age of sixty-nine. The lecture, "On Mr. De la Rue's Photographic Eclipse Results," was held on May 3, 1861. Unlike Airy's 1851 eclipse preview or Smyth's firsthand expedition account, Faraday recounted from a third-person perspective the observation of a solar eclipse that had occurred the previous year. He introduced the eclipse photographic works made by De la Rue's new instrument, the Kew Photoheliograph, during an eclipse expedition to Spain.[29] Unlike Smyth's more general travel narratives, Faraday's lecture contained substantial scientific information and technical details. Faraday started by explaining the principles of solar eclipses, then proceeding with the design of the instrument, the phenomena of prominence, corona, and sunspots, and showing De la Rue's photographic results. As a highly respected senior man of science, Faraday unsurprisingly attracted an extraordinarily large audience, yet Smyth was no less popular. The two lectures were very well attended: 521 people at Smyth's lecture and 783 at Faraday's, when the average size of Friday Discourse audience was only about 470 in both years.[30]

It may not seem surprising that lectures at venues like the Royal Institution, where many practitioners of science participated, presented stories of scientific discoveries or the latest progress in the discipline. Private entrepreneurs' lectures, however, could also carry messages of scientific news and novelties. Lecturers were especially eager to advertise that their mechanical apparatus exhibited in the show was the most innovative or improved. In an 1851 advertisement, C. H. Adams, for example, assured the audience that the "nine new planets"—possibly the newly discovered asteroids and Neptune—had been added to his orrery.[31] *Punch* magazine had poked fun at Adams's regular update of his orrery. The *Punch* journalist claimed that for Adams's sake astronomers should "give him [Adams] breathing-time," as discoveries were now "going on so rapidly, that it costs him a little mint per annum in adding new planets."[32] Recent or upcoming celestial spectacles such as eclipses, comets, and meteor showers, could be sensational stories in the newspapers and were hardly ignored by lecturers. A competent lecturer had to be aware of the major news and current events in scientific communities. George Bachhoffner, for example, organized a series of daily lectures on the solar eclipse at the Colosseum in 1860. This series of daily lectures began at least ten days before the eclipse, which occurred on July 18, 1860, and was a special arrangement for this sensational spectacle.[33]

A rare poster of C. H. Adams's lecture preserved in the Victoria and Albert Museum, London, shows another example of the way that popular lectures echoed a scientific news headline. This case is related to the discovery of the hypothetical planet Vulcan, one of the most intriguing mysteries in the history of astronomy in the nineteenth century. The existence of Vulcan between Mercury and the Sun had long been a subject of debate among astronomers. Vulcan was the El Dorado of astronomers. The story of the race to find this hypothetical planet was fascinating because it involved national pride and the triumph of scientific reason, and it was an inspirational legend about the sudden fame of an underdog.

The story must begin with the anomaly of Mercury's orbit. The perihelion of Mercury's orbit has a slight advance—43 arc seconds per century—making it inconsistent with theoretical calculations. This puzzle, known as perturbations, had bothered astronomers for a long time. In 1859 Urbain Le Verrier, the director of the Paris Observatory and the *grand homme* powerfully dominating the direction of astronomical research in France, set out to tackle this unsolved problem. As a virtuoso on Newtonian mechanics, Le Verrier, whose calculations led to the discovery of Neptune in 1846, tried to repeat his success in finding the new planet. The cause of Mercury's perturbations, he surmised, was a hitherto unknown mass, either a planet or a group of asteroids, standing between Mercury and the Sun. Le Verrier spelled out this hypothesis in a letter published in the journal of the French Academy of Sciences on September 12, 1859. He launched an appeal to astronomers to seek out this unknown planet.

Issued by the most powerful person in French astronomical science, this appeal could not be ignored, especially coming from the man who discovered Neptune about a decade earlier. Le Verrier's hypothesis stirred great interest among astronomers all over the world. His letter was translated into English and reprinted in the November issue of *Monthly Notices of the Royal Astronomical Society* in the same year, causing alarm in Britain, for it was a forcible reminder of "the period preceding the discovery of Neptune," as noted by the journal's editor.[34] The British, who had lost in the race to discover Neptune, could not accept that they had been bested by their French rivals.

Things took a dramatic turn toward the end of 1859. Le Verrier received a letter from a fellow Frenchman, Edmond Lescarbault, a country physician and an unknown amateur astronomer, claiming that he had already *seen* the hypothetical planet from his private observatory in Orgères-en-Beauce in March of the same year. Due to his diffidence, Lescarbault had not announced his result immediately, instead waiting until he read a journal article on Le Verrier's work. Lescarbault's procrastination was partly due to his being an unknown amateur. It annoyed Le Verrier. To verify the claim, Le Verrier made a surprise visit to Lescarbault before New Year's Eve. Details of the meeting were made known in an article by the celebrated science writ-

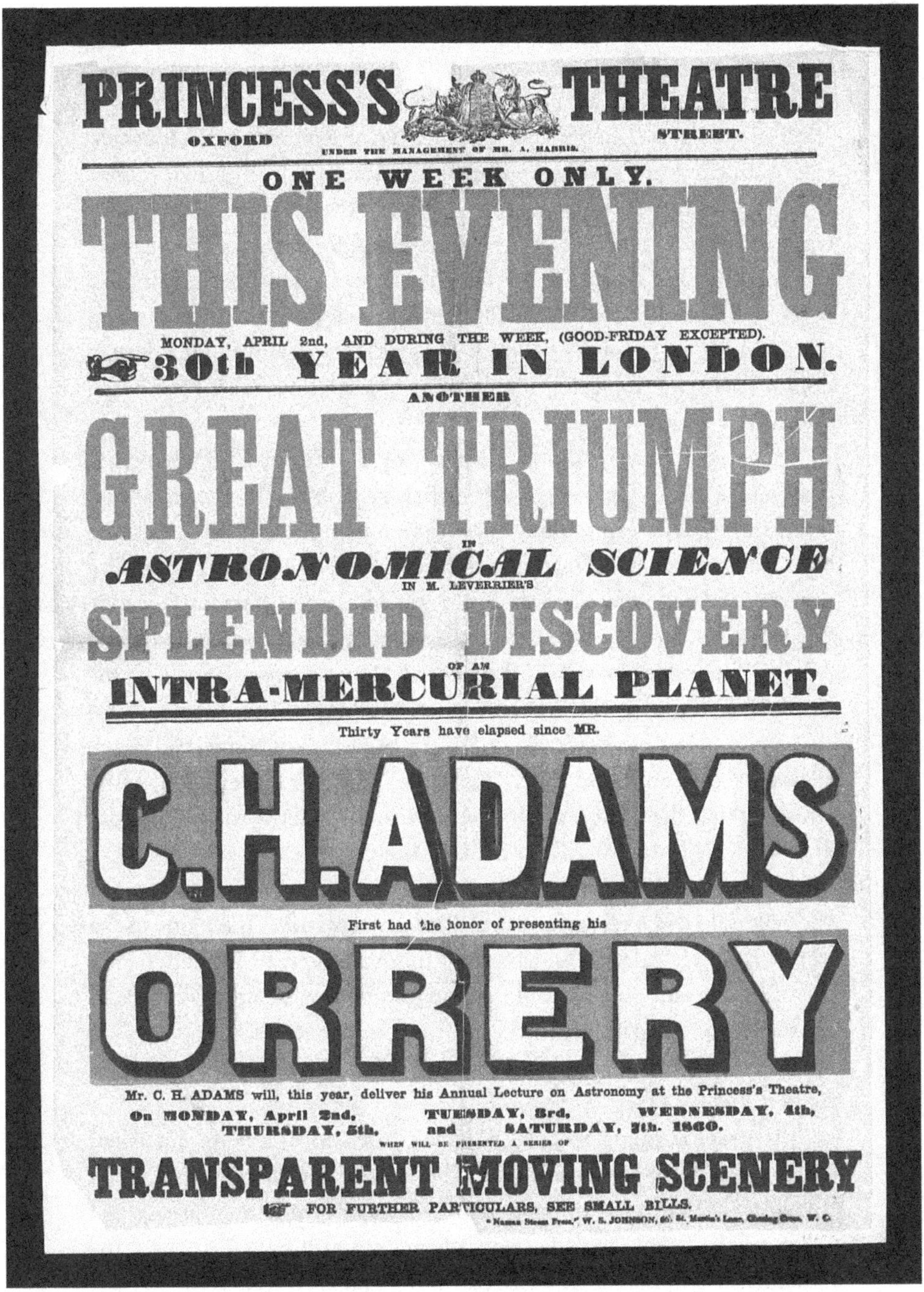

FIGURE 4.1. Poster of C. H. Adams's lecture featuring the discovery of an "Intra-Mercurial planet" at the Princess's Theatre, April 2, 1860. V&A: S.1702-1995. © Victoria and Albert Museum, London.

er Abbé Moigno in the French journal *Cosmos*, which was later translated and widely cited in the English press. Moigno amusingly described the drama of a country "lamb" unexpectedly encountering a "lion" from Paris. The poor physician, unaware of the guest's identity, was "subjected to a severe cross-examination by his unknown visitor, who pressed him hard from step to step."[35] Eventually, Lescarbault's testimony convinced Le Verrier. Returning to Paris, Le Verrier announced the discovery at the French Academy of Sciences on January 2, 1860. The newly discovered planet was named Vulcan after the Roman god of fire. Lescarbault suddenly rose from an unknown country doctor to a scientific celebrity and was made Chevalier de la Légion d'honneur within a month.

By early February 1860, the news of the discovery had spread, receiving extensive coverage in the British press.[36] The inner planet within Mercury's orbit, along with the amusing tale of the dramatic encounter between Le Verrier and Lescarbault, became an immediate sensation. Newspaper readers could not miss such a scientific headline story, and neither did seasoned astronomical showmen like C. H. Adams. He made the newly discovered planet a main feature of his April 1860 lecture. The lecture's poster demonstrates Adams's precise grasp of the big news (fig. 4.1).[37] In bold type it shows the venue (Princess's Theatre) and time ("One week only. This evening"), while stressing that the season marks the lecture's thirtieth year in London. The headline in the middle section of the poster catches the viewer's eye: "Another great triumph in astronomical science in M. Leverrier's [sic] splendid discovery of an Intra-Mercurial planet." The name of the lecturer, C. H. Adams, and the iconic instrument in his repertoire, the orrery, is embossed in bold. The organization of the poster would cause a passerby to notice the keywords at a glance. Its striking design combines the sensational headline story with the lecturer's long-term reputation.

The exact content of C. H. Adams's April 1860 lecture, however, is not clear. Very few accounts of the lecture are known aside from this poster. The literary magazine *Athenaeum* briefly mentioned it in a summary of that year's Passion Week entertainments. As the journalist described it: "At the Princess's Mr. Adams exhibited his Orrery, and delivered his usual Astronomical Lecture, including among its topics M. Leverrier's [sic] discovery of an intramercurial planet."[38] In modern hindsight, the existence of Vulcan was never truly verified and skeptical voices persisted in the next few decades. Despite consistently unsuccessful later observations, Le Verrier kept his faith in Vulcan to the end of his life in 1877. The elusive intramercurial planet remained an unsolved mystery for the rest of the nineteenth century, until Albert Einstein's explanation of Mercury's anomaly using his theory of general relativity in the late 1910s "destroyed" the planet.[39] The case of C. H. Adams's April 1860 lecture nevertheless suggests that private astronomical lectures not only could provide familiar textbook knowledge but also contain fresh scientific news. Apart from the Sun, the Moon, the tides, and Newtonian science, there were wider possibilities of topics that popular astronomy lectures could provide. Stories of the discovery of Vulcan remind us of the roles played in the construction and circulation of knowledge by often "invisible" contributors other than elite scientists, such as Abbé Moigno the writer and C. H. Adams the lecturer.

Although both astronomical lectures at scientific institutions and those presented by private entrepreneurs could contain topics of the latest scientific progress, the two types differed. Historians of science have examined nineteenth-century cases of the relationships between astronomical communities and the mass media. Much research suggests that there was a symbiosis between astronomers and public coverage of astronomical events—the popular press did not simply follow scientific events or progress passively. Rather, they often collaborated with or were used by scientific practitioners to actively influence public opinion, thereby affecting actual scientific work. During the preparations of the 1874 transit of the Venus British expedition, for instance, Richard Anthony Proctor, the editor of the *Monthly Notices of the Royal Astronomical Society*, used the journal and other newspapers as channels to challenge the plan of George Airy and the Admiralty, who were in charge of the expedition. Subsequently, during the craze for Martian astronomy in the late nineteenth and early twentieth centuries, astronomers used newspaper coverage to attract public attention while also reshaping the norms of practice within the discipline.[40] In the above cases, as Jessica Ratcliff indicates, science journalism in this period was "not easily separable from the working world of science." Science journalism was often an arena in which the disciplinary dispute internal to the astronomical community was "put on display to the public in the hope of attracting political support," rather than simply transmitting scientific ideas to the public for educational purposes.[41]

Ratcliff's remark also applies to the relationships between the popular press and astronomical lectures at a scientific institution, especially an elite organization like the Royal Institution, as the lectures were often given by men of science or working astronomers. Their lecture contents were more likely related to original research or scientific work in progress, thereby serving internal scientific purposes or eliciting external public support. Airy's Friday Discourse prior to the 1851 total solar eclipse, for example, was to some degree a public relations campaign enhanced by his offer of a free guide issued by a professional committee, and aimed at showcasing the scientific work of the Astronomer Royal and his colleagues. These firsthand reports sometimes contained considerable technical detail, and the "news" or "topical" nature of the lectures lay in science rather than in the event's popularity in news coverage. The Royal Institution Friday Discourse series reflected this tendency well. For example, Airy's lecture on "The Pendulum-experiments lately made in the Harton Colliery, for ascertaining the mean Density of the Earth" in 1855 and Baden Powell's "On Rotatory Stability; and its Applications to Astronomical Observations on board Ships" in 1858, were quite technical in content and could hardly appeal to a general audience. Even the more absorbing talk about the Tenerife expedition by Smyth, without his adjustment to make the story accessible, was originally not expected to draw the attention of the public. Smyth admitted that interest in the mission of the Tenerife expedition was initially shown "in the limited circle of working astronomers." This experimental expedition was therefore not a headline story in the first place despite its scientific value.

In contrast, a private astronomical lecture such as D. F. Walker's or C. H. Adams's in a theater reflected more popular taste and was more likely to follow the hottest headlines. As these private entrepreneurs were not usually involved in astronomical research, they were expositors of science rather than original investigators. Their lectures had little impact on actual scientific practice. The popular press was a source inspiring their lecture topics, rather than a tool to influence scientific work. Compared to the technical details of the latest scientific experiment, a recent celestial spectacle or a sensational astronomical discovery featured in the newspapers was more likely to be incorporated into a private lecturer's program.

A Playwright's Work

Entrepreneurs of private astronomy lecturing often drew inspiration from other popular works. Samuel James Arnold, the author and producer of George Bartley's lecture, acknowledged that he applied other sources in the manuscript submitted to the Lord Chamberlain's Office. "So many excellent works having been written on this subject it can hardly be expected that any striking novelty either in language or arrangement should be attempted ... and flights of imagination would be strangely wasted on a Theme so vast, that the clearest intellect becomes bewildered in the contemplation of its immensity." Arnold then explained his methods: "To confine ourselves therefore to what is known.... In so doing we shall draw largely on the works of the best Authors who have written on the subject, because nothing can be added to that which is already complete and full."[42] Although Arnold did not specify which sources he relied on to write the lecture, the manuscript he submitted to the Lord Chamberlain shed light on how the knowledge and styles expressed in popular astronomy lectures in the early nineteenth century were circulated, copied, and disseminated.

This manuscript, titled *Ouranologia*, was a detailed syllabus of Bartley's astronomical lecture presented to the Lord Chamberlain's Office in 1826. It is uncertain, however, why Arnold would submit it to the Lord Chamberlain. Although the Lord Chamberlain held the power to examine a new drama before public performance, this censorship did not apply to Lenten amusements such as Bartley's lecture in the English Opera House.[43] No other astronomical lectures in theaters are known to have been submitted to the Lord Chamberlain's Office. Besides, even if Bartley's lecture was required to be examined, it was not a new show waiting to be licensed. It had been performed in the past few years. In any case, there seems no need for such a submission. The Lord Chamberlain's Office register of plays simply notes the title of the lecture and the theater in which it was performed without further information.[44] Perhaps the submission was at the Lord Chamberlain's personal request, since Arnold wrote in the covering letter that he was "in obedience to the instructions I have received at your Lordship's office."[45] Various other reasons were possible: Arnold probably submitted the manuscript on his own initiative to impress the Lord Chamberlain, to secure some form of copyright protection over the show's content, or to avoid the controversy of using a well-known actor (Bartley) for the "performance" during Lenten restrictions. The last guess was particularly likely, considering that Bartley's show was the only known Lenten astronomy lecture produced by a theater manager and performed by a renowned comedian in a West End theater. Bartley and Arnold might have faced pressure or complaints from other theater rivals.[46] We will never know which speculation, if any, is true unless further evidence emerges. Nevertheless, Arnold's submission of the lecture manuscript unexpectedly provided a fine specimen of Lenten astronomy lecture in the early nineteenth century. This manuscript is now preserved in the Lord Chamberlain's plays collection at the British Library.

Samuel James Arnold was a theater celebrity with a keen interest in science and literature. He was the eldest son of Samuel Arnold, a renowned composer and organist at Westminster Abbey, who had a particular interest in oratorio. This influential background offered Samuel James Arnold significant resources during his early career. The father and son worked in close collaboration: many musical pieces in Arnold's early plays were composed by his father. Arnold junior soon pursued a theater career as a playwright. In addition to writing plays, he was also actively involved in theater management. He obtained a license to change the old Lyceum Theatre into an English opera house. Meanwhile, he was also invited to become a joint manager of the Drury Lane Theatre. However, his management in Drury Lane was troubled by internal arguments and he eventually resigned in 1815 to concentrate on the redevelopment of the Lyceum. The rebuilt English Opera House opened in 1816, and Arnold stayed on as its proprietor and manager for the next two decades. Besides his professional career in the theater, Arnold was also a science enthusiast. He was involved in the activities of the Royal Institution, including its management. His name appeared on the balloting list for the committee of General Science, Literature, and the Arts at the Royal Institution at least twice, in 1810 and 1811, respectively.[47]

D. F. Walker's successful exhibition of the eidouranion at the English Opera House in 1817, as I have shown in chapter 2, probably inspired Arnold to produce his own Lenten astronomy show. A successful lecture required a lecturer with great showmanship skills, so Arnold turned to the seasoned comedian George Bartley. The two had already become acquainted before the show: Bartley and his wife Sarah played at Drury Lane in 1814, when Arnold was still the comanager there. After the married couple returned from a successful tour to the United States in 1818, they settled in London, and Bartley accepted new stage engagements at Covent Garden in winter and at the English Opera House in summer. It was also then that Bartley and Arnold started collaborating on Lenten astronomical lectures at the English Opera House.

Although Arnold acknowledged that he drew largely from popular works of the best authors on astronomy when writing the lecture syllabus, he was not explicit about which sources he adopted. The name of the lecture "Ouranologia" was the same as a small book printed a century before in 1695 by Thomas Cole, but the two works had nothing in common but their titles. *Ouranologia* (1695) by Cole was an ephemeris for astrological use and included a collection of articles on astrology, while Arnold's work was a substantial summary of the rudiments of astronomy and had no relation to astrology. The similarity of the titles was just a coincidence: the word "ouranologia" is derived from *Ouranos*, an ancient Greek word meaning sky or heaven, and the suffix *logia* from *lógos* meaning explanation, discourse, or reason. This usage was similar to the purpose for creating the words "eidouranion" or "dioastrodoxon." Eighteenth- and early nineteenth-century contemporaries were fond of inventing fancy compound words with classical roots to display their learnedness.

The arrangement of subjects in Arnold's *Ouranologia* resembled the structure of the book *Astronomy Explained upon Sir Isaac Newton's Principles* (1756) by James Ferguson. Ferguson was a prominent astronomy lecturer and author in the mid eighteenth century (see chapter 1). His *Astronomy* was reprinted many times and was even republished posthumously. For instance, it was published in Philadelphia in 1809 as the second American edition, with revisions and corrections made by the professor of mathematics Robert Patterson at the University of Pennsylvania. David Brewster also edited and republished it in Edinburgh with supplementary chapters in 1811. This popular astronomical classic remained influential even more than half a century after its first publication, so it was definitely not "outdated" in Arnold's time.

Judging by the structure, quotations, and almost identical opening paragraph of the introduction, though Arnold did not refer specifically to Ferguson's *Astronomy*, this classic was very likely the main reference for the *Ouranologia*. In addition to Ferguson's work, Arnold also quoted from popular books such as *An Introduction to Astronomy* (1786) by John Bonnycastle. Nevertheless, Arnold did not simply copy the above works. He still made a huge effort to paraphrase the texts and incorporate them into different scenes of his own astronomy lecture. The example of the *Ouranologia* illustrates the way that popular astronomy lecturers of the early nineteenth century imitated and quoted bestselling astronomy textbooks from the late eighteenth century.

Arnold divided the lecture into three parts. The first part shows the shape of the Earth using the "Diagram of Ship" to demonstrate the curve of its surface.[48] It also includes different cosmological models in history: Pythagorean, Ptolemaic, Tychonic, and the "genuine one of Copernicus." The stories of historic philosophers and astronomers finally lead to the achievement of Isaac Newton, who "established the Copernican System upon such an everlasting basis of mathematical demonstration, as can never be shaken."[49]

The second part presents an overview of the solar system, emphasizing the telescopic appearances of individual planets. Comets constituted another absorbing subject: three scenes of the comets of 1680, 1811, and 1819 are included. Arnold stressed that the comet in each scene was "laid down from actual observation when it was in that part of the Heavens." The scenes were thus faithful representations of the comets' appearance in the constellations as they were situated at the time.[50] The Sun and the Moon are also elaborated on, with many extra scenes such as the Sun's apparent magnitude to the different planets, and the orbit and phases of the Moon.

The third part focuses on the eclipses and the tides. The former were especially phenomena described as "none that engage the attention of Mankind more . . . nothing appears more extraordinary than the accuracy with which they can be predicted."[51] The causes and different types of eclipses are explained in detail, as are the tides. The lecture concludes with the display of the extensive orrery, which contains the whole solar system and every planetary movement.

Like many other public lectures on natural philosophy in the eighteenth and early nineteenth centuries, the significance of the legacy of Newtonian science is obvious in the *Ouranologia*. Newton was the main protagonist among a long line of great geniuses. Arnold's high praise for Newton reflected the prestigious status of this English national hero. "All other systems have been wholly exploded by the clear and demonstrative discoveries of our immortal Countryman Sir Isaac Newton," Arnold concluded in the first part of his lecture, "who has shewn that though ingenious Argument might suppose the course of nature to be governed by mere mechanical laws only, the works of nature would then have been incomparably inferior to what they now are both in beauty and perfection."[52] The phenomenon of the tides occupied such a large part of the *Ouranologia* because the subject best exemplified the Newtonian principles of gravitation.

The text of the *Ouranologia* was more than a mere script or a common syllabus. It was neither made for the lecturer to read every line, nor intended simply to list the title of topics in the lecture. The manuscript more resembled a copy of "production notes" with the playwright's thoughts on what effects the audience could expect to see. It was also like an outline of the lecture written to justify what was to be performed onstage. In the manuscript, Arnold clearly explained his objective and the theme of the lecture. Sometimes he even explained why he did or did not adopt a scene in a particular part. For example, when introducing the change of lunar phases, he commented that the recurring progress "could shew on this apparatus, but as it would only reverse the succession of the same forms which have just been shewn[,] it might be considered as an unnecessary waste of time: particularly as the nearest Scene will shew in a different manner."[53]

Another example in the *Ouranologia* showing the playwright's considerations was the "Diagram of Ship" scene. In this scene, a model vessel would appear on the top of the globe to demonstrate the spherical shape of the Earth. Because of the different heights at which spectators were seated in the auditorium, this effect could vary. Arnold had noticed such spatial differences, and thus he explained: "A Ship will shortly appear on its [the globe's] surface, advancing towards the Audience—Those of the Spectators who are situated in the higher parts of the Theatre, will of course behold it first—As a Seaman in the foretop first discovers a Sail at Sea—Those persons in the lower parts of the Theatre will perceive it later—but I trust all of my Auditors who favor me with their attention, will find that its advances are precisely correspondent with my description; thus illustrating the Globular shape of the Earth."⁵⁴ This remark exactly shows a subtle technical detail considered by the producer, and how reliable the showmanship of the lecturer was for the performance. Arnold believed Bartley could easily distract the audience from this slight difference due to the unavoidable restrictions of the auditorium space. This also indicates Bartley's importance as the lecturer. Though Arnold was the sole author of the *Ouranologia*, Bartley's role as the showman executing the program should not be neglected.

Devotion, "Daughter of Astronomy"

The *Ouranologia* also shows a significant theme that was often addressed by popular astronomy lecturers in early nineteenth-century Britain: the wonder of the universe that displays God's creation. Religion played an important part in popular science discourse, and astronomy was no exception. The religious tone was prevalent and proper for narrating the vastness and order of the heavens. Nothing could be better than the infinite universe to elicit feelings of awe and religiosity in the audience; nothing could be better to enhance the public's interest in astronomy as the sublime science. As Allan Chapman has observed, the promotion of a cause was an important ingredient in nineteenth-century astronomical lecturing, and the most common single cause promoted by lecturers was that of Christianity itself.⁵⁵

A good example reflecting the association between astronomy and religious devotion is a frequently used quotation: "An undevout Astronomer is mad." This quote appears at the very beginning of Ferguson's *Astronomy*, where the author notes that he is citing Dr. Young's *Night-Thoughts*. Edward Young, an English poet and a contemporary of James Ferguson, is best known for his long poem *The Complaint: or, Night-Thoughts on Life, Death, & Immortality*, commonly known as *Night-Thoughts*. The line Ferguson quoted is from the following passage:

> Devotion! daughter of Astronomy!
> An undevout astronomer is mad.
> True; all things speak a God; but in the small,
> Men trace out Him; in great, He seizes man;
> Seizes, and elevates, and wraps, and fills

> With new inquiries, 'mid associates new.
> Tell me, ye stars! ye planets! tell me, all
> Ye starr'd, and planeted, inhabitants! what is it?
> What are these sons of wonder? say, proud arch
> (Within those azure palaces they dwell),
> Built with divine ambition! in disdain
> Of limit built! built in the taste of heaven!
> Vast concave! ample dome! wast thou design'd
> A meet apartment for the Deity?—[56]

Ferguson thought that the line "An undevout Astronomer is mad" was too obvious to be hyperbolic. In his book, Ferguson emphasized the moral objective of the study of astronomy, and he expected that readers could learn "by what means or laws the Almighty carries on, and continues, the admirable harmony, order, and connexion observable throughout the planetary system."[57]

As a follower of Ferguson's popular work, Arnold also used this quotation in the *Ouranologia*. He wrote at the beginning of the syllabus: "Thus Astronomy becomes a handmaid to Devotion, and affords us the most exalted ideas of that beneficent Deity who created, guided, and governed, the stupendous whole in matchless harmony. Well has it been said that 'the undevout Astronomer is mad.'"[58] An inexpensive book, *The Solar System* (1799), was published by the evangelical Religious Tract Society and aimed at a relatively elementary readership. It also quoted this line in its introduction and declared that astronomy "ought to be considered as bearing an intimate relation to religion, and worthy the study [sic] of every enlightened Christian."[59] This "undevout astronomer" quotation makes a perfect footnote for a popular attitude toward the study of astronomy in the eighteenth and early nineteenth centuries. On the one hand, astronomers' stargazing practice is benign as it leads to a religiously beneficial result; on the other hand, an undevout astronomer must be insane as the ignorant stargazer cannot see the manifest display of the design from an intelligent creator.[60]

Not only lecturers themselves stressed the connection between astronomy and religious devotion; patrons also often expected to obtain this profitable cultivation of religious sense from lectures. An anecdote about Rev. John Eyton, the Vicar of Wellington, Shropshire, best exemplifies such religious and moral expectations. A visiting lecturer of

FIGURE 4.2. "Order Is Heaven's First Law," illustration on the title page of the vignette, *Real or Constitutional House That Jack Built*, published by J. Asperne, London, 1819. Image digitized by HathiTrust.

astronomy received permission to give a course in a local free school. Upon the lecturer's arrival, he attended the evening church service, chaired by the vicar, to advertise his lecture. To make the upcoming lecture "really profitable to the young persons of his charge," and "to prepare them for a right understanding of the sublime science," Rev. Eyton preached on Psalm 8:3, "When I consider thy heavens," and addressed astronomy and the greater wonders of redemption at length to the audience. The beleaguered visitor had not expected the vicar to preach about astronomy and was "little prepared to hear."[61]

Like the "undevout astronomer" quotation, this anecdote of Rev. Eyton implies that astronomy is a double-edged sword: it could inspire religious awe but could also undermine the basis of Christianity. The idea of a clockwork universe, which proposes that nature operates in an orderly fashion according to mechanical principles, had long fascinated European philosophers during the Enlightenment. Newton's mathematical description of the force driving planetary motion made a mechanical universe more plausible. Although Newton himself affirmed God's hand in the natural laws he discovered, the physical laws of gravitation could be seen as driving the operation of the universe independently of divine activities. A purely mechanical universe, as the French scholar Pierre-Simon Laplace's work *Mécanique céleste* suggested, is a materialist system in which God and moral dimensions were no longer needed.[62] Political turmoil after the French Revolution complicated this fear of subversive radicalism. Such connections with materialism and revolution worried many British contemporaries. This sentiment was best represented in the frontispiece of a vignette satire *The Real or Constitutional House That Jack Built* (1819).[63] In this allegorical illustration (fig. 4.2), the crown, the constitutional principles, the English laws, the Bible, and a drawing of the planetary system are laid on a solid rock, symbolizing the social stability constructed upon these things. In contrast, the house on the other side of the Channel crumbles, implying the devastation caused by the French Revolution. The picture of the solar system is marked with an inscription "Order is heaven's first law." The entire image cleverly expresses the connections between politics and astronomy. Astronomy could be benign, but careful guidance is required for the right understanding of the sublime science, as Rev. Eyton was doing. Edward Young's *Night-Thoughts* was also full of reflections on the universe and the Creator, including that the book of stars is universally available and reveals God's existence and nature. Therefore, an undevout astronomer is not only mad but also dangerous. Materialism, atheism, and revolution were all dangerous notions and ought not to be the daughters of astronomy. This is why many nineteenth-century British contemporaries thought religious reflections, particularly the indication of the Creator, were necessary in astronomy lectures: they functioned to make the science not only sublime but also safe.

The idea and rhetoric of natural theology was especially crucial to popular science across various Christian denominations. William Paley's *Natural Theology* (1802) was extremely popular and influential in nineteenth-century Britain. It presents many teleological and cosmological arguments for the existence of God and the Creation.[64] The most iconic design argument used by Paley in *Natural Theology* was the watchmaker analogy: a fine and self-running watch must have been contrived by a watchmaker rather than randomly by nature. Paley extended this analogy to argue that complex structures of living beings also require an intelligent designer. Although astronomy seems not as directly relevant to the watchmaker analogy as biology or physiology, the orderly revolution of the planetary system also implies the mighty power of the Creator. Paley, despite rejecting astronomy as the best medium through which to prove the agency of an intelligent designer, admits that astronomy shows the Creator's magnificence better than all other branches of science.[65] Other later natural theologians, such as William Whewell and Baden Powell, had views different from Paley's on the theological significance of the order of the universe, but they would agree with Paley's judgment about the advantage of astronomy as an incomparable source of displays of the sublime.

The utility of natural theology in popular astronomy lecturing for most lecturers therefore drew on language that evoked emotion rather than rigorous theological arguments. The rhetoric of natural theology, particularly the implications of the design argument, was widely used by lecturers and authors to convey the sense of awe and wonder elicited by the Creation of a benevolent God. Biblical or literary works, such as Psalm 19:1, "The heavens declare the glory of God; and the firmament shewth his handywork," and John Milton's poem *Paradise Lost*, were frequently quoted in popular lectures, including the *Ouranologia*. Bartley's lecture was not an isolated example. Many other lecturers were also in tune with the emphasis on religious significance. For instance, R. E. Lloyd advertised that his astronomy lecture was intended to direct the inquiring mind through Nature, "up to Nature's God," and therefore "Seminaries and Families" will find his lecture "peculiarly interesting and grateful."[66] D. F. Walker also advertised his objective and endeavor in playbills, claiming that he aimed to elucidate "the sublime Science of Astronomy, on a Scale commensurate to its Importance, to imitate, though humbly, the glorious Phenomena of Creation."[67]

John Brooke and Geoffrey Cantor have analyzed seven rhetorical devices used by natural theologians: vivacity, beauty, sublimity, novelty, analogy, metaphor, and antithesis. Especially the first four are qualities that stimulate imagination and "gratify the fancy."[68] Early nineteenth-century popular astronomy lecturers often applied these rhetoric devices, particularly sublimity and novelty. The former is obvious, as the universe is an unparalleled source of inspiration for feelings of the sublime. The latter manifests itself in a strategy of drawing attention to extraordinary facts about the physical world, such as comparisons between the size of the Sun and the Earth (9 feet in diameter versus a globe of only 1 inch), the length of a comet's tail (30 million miles) and the number of visible stars in the Milky Way (116,000 in the most crowded part; 90 million through the largest telescope).[69] This strategy is not limited to static numerals but sometimes even applies to a dynamic picture. For example, in his volume of the Bridgewater Treatise Whewell drew an analogy between small marble pellets in a very large basin and planets in the Solar System, to shape a vivid image combining an extraordinary universe and familiar objects.[70] These physical facts generate a sense of novelty and further extend to a coherent and profound meaning: reverence for the immeasurable wisdom and power of God and the order he created.

Religious renditions of the Creation and design were evident in the *Ouranologia*. Arnold repeatedly reminded the audience about the advantage of astronomy: a sublime subject affording the knowledge of nature, the true system of the world, and the invariable laws by which it is governed. He claimed that astronomy "has opened to us such a magnificent view of the Creation, that we are struck with astonishment at the grandeur of the spectacle, and the powers of omnipotence."[71] When introducing the cause of the seasons, Arnold attributed this wonderful mechanism to the benevolence of the Creator:

> This beneficent and curious provision for the existance [sic] and comfort of the Earth's inhabitations cannot too powerfully excite our admiration of the wisdom, or our gratitude for the goodness of the Creator. If it were not for this simple contrivance one part of the Globe would revolve constantly in the full blaze of the Sun's rays—while those regions which are situated near the poles would be almost, wholly, destitute of light and heat, and probably incapable of sustaining either Animal or vegetable life.—But this is not the case, for the remotest points to which the avarice or curiousity [sic] of Man has penetrated are sound to be inhabited; and doubtless that power which "tempers the wind to the shorn Lamb"—has so organized their inhabitants as to afford even in those desolate regions the Capability and means of enjoyment.[72]

The narrative of these passages fitted the rhetoric of natural theology. By using astronomical knowledge, natural theologians argued that these physical facts possessed profound significance and showed the order created by a benevolent designer. Whether Arnold intended to advocate this natural theological idea or he unintentionally followed this fashion in popular scientific publications, he was neither the first nor the last author to do so. The *Ouranologia* exactly reflects how influential such religious narratives had been in the heyday of natural theology.

Since Lloyd, Walker, and Bartley were all familiar names in theaters during Lent, it is tempting to think that the tone of religious reflection in popular astronomy lectures was exclusively a Lenten feature. However, subjects related to natural theology were not limited to Lent or staged astronomy shows. Some astronomical lectures taking place in scientific institutions beyond Lent shared similar religious flavor. Baden Powell's course of seven lectures on "Cosmical Philosophy" at the Royal Institution in 1851, for example, addressed many theological or philosophical subjects in the syllabus. Powell indicated the difference between his present course and his previous one, "Astronomy," in the preceding year. The previous course (astronomy) referred to "the phenomena and laws constituting the system of the world," and the present one on cosmic philosophy relates to the "investigation of their causes and the general philosophical principles involved." This course was a rendition of the history of astronomy in general. In the final lecture, Powell elaborated philosophical topics highly relevant to theology, such as "Better distinction into physical and moral," "Evidence of design" and "Universality of order."[73]

Another example was a course of three lectures on "Ancient and Scriptural Astronomy" at the Royal Manchester Institution, delivered by the Rev. St. Vincent Beechey in 1850. This course included many biblically interpretive topics.[74] Like Powell, Beechey made a similar distinction between religious connotations and physical knowledge in astronomy. In another course, "History of Modern Astronomy" at the Royal Manchester Institution, Beechey plainly addressed the development of astronomy from the ancient Greeks to the recent discovery of Neptune without any rendition of biblical context. On the one hand, this kind of demarcation as adopted by Powell and Beechey could suggest that a division between physics-based astronomical subjects and theology-oriented philosophical discourses had reached a consensus in institutional lecturing by the mid-nineteenth century. On the other hand, it could be interpreted as a reconciliation rather than a division: science was commensurate with and justifies Christianity. To talk about the mere scientific facts of the physical world was not enough; religious philosophy ought to be a supplement or a guide to complete the sciences, hence this kind of arrangement. The lecturers, Powell and Beechey, were pleased to arrange two respective courses on physical and theological worlds with a short interval.

Progress and the Plurality of Worlds

As I have indicated, the discourse of natural theology was influential in popular astronomy lectures, but its strength was in the rhetoric and emotion rather than in provision of rigorous theological arguments. The reason was partly because natural theologians could not reach a consensus on some disputed issues. The notions of progress and the plurality of worlds, for example, were two important core issues in contemporary astronomical debates linked to natural theology. The former centered particularly on the "nebular hypothesis," while the latter implied the existence of extraterrestrial worlds and beings. Unlike the firm status of Newtonian laws of gravity, both ideas could be disputed and cause heated scientific and theological controversy. There was, however, a contrast between the two ideas in their acceptance by the public: the nebular hypothesis with its notion of progress was less favored, but the idea of the plurality of worlds was supported by most astronomers and critics in this era. This difference caused the former to be almost neglected in the syllabi of astronomy lectures, while the latter was a popular topic among some astronomical lecturers.

Progress, or the "science of progress" as Simon Schaffer calls it, was a central debate topic among scientific elites in early Victorian Britain. It is straightforward as well as difficult to define what "progress" is. Progress means that things change and develop into a current state through a certain process and will continue developing. Nineteenth-century contemporaries often asserted that progressive principles applied not only to nature but also to human affairs such as society and civilization. The historian J. B. Bury explains that human progress is "a theory which involves a synthesis of the past and the prophecy of the future. It is based on interpretation of history which regards men as slowly advancing ... in a definite and desirable direction."[75] The notions and narratives of progress diffused into various scientific disciplines, especially geology and astronomy.[76]

Geological findings of rock strata and fossils beginning in the seventeenth century challenged the conventional Genesis story. Geological evidence inspired savants across Europe to reach fresh insights into the Earth's structure and history; their theories could provide an alternative scope for the progressive history of the Earth and life. In this alternative narrative, conventional biblical events such as the Creation and the Flood seemed unable to fit the recently extended enormous timescale of the Earth.[77] Geologists, especially those who were also clerics such as William Buckland, carefully handled these progressive issues and tried to reconcile science and theological exegesis. In the view of these cleric geologists, the geology of progress was harmless to Christian belief and the story of progress could be divinely directed to "prepare the earth to humans."[78] The caution of geologists was not without reason. Progress and the corresponding cosmogony were "consciously fashioned tool[s] with distinct persuasive purposes."[79] The science of progress was easily connected to the evolutionary debate, radical reform, and revolution. This dangerous connection had been invoked long before the anonymous publication of Robert Chambers's *Vestiges* (1844) or Charles Darwin's *Origin of Species* (1859). For the proponents of progress, the image of universal progressive development was reflected not only on the Earth but also in human society. The science of progress rightly endorsed their visions based on rock formations and the celestial nebulae.[80]

In the realm of astronomy, the nebular hypothesis was the main progressive discourse used to challenge the biblical narrative. Although its roots can be traced to Immanuel Kant, Laplace, and William Herschel, the term "nebular hypothesis" was not coined until Whewell's Bridgewater Treatise.[81] The nebular hypothesis proposes a mechanical scenario for the origin of the solar system: it started with a mass of gaseous nebula. Through the action of natural laws, this condensing and rotating nebula contracted, and its matter precipitated into separate rings. These rings of matter eventually formed the Sun, planets, moons, and all other celestial objects. Whewell described this theory in his Bridgewater Treatise on astronomy and general physics in 1833. Like those geologists in their efforts to accommodate science and religion, Whewell interpreted the hypothesis as a divine cause and effectively refused to countenance the dangerous materialism associated with Laplacian cosmogony. He suggested leaving the scientific merits of the nebular hypothesis to be judged by future generations. Regardless of whether it was scientifically correct or wrong, he asserted that "the final fate of this opinion cannot, in sound reason, affect at all the view which we have been endeavouring to illustrate;—the view of the universe as the work of a wise and good Creator."[82]

Whewell's Bridgewater Treatise brought the nebular hypothesis to the attention of a popular readership, but the hypothesis became even more prominent when the Scottish astronomer and political economist John Pringle Nichol advocated a point of view contrary to Whewell's.[83] Nichol's popular book *Views of the Architecture of the Heavens* (1837) enthusiastically promoted the nebular hypothesis. Nichol's endorsement of the nebular hypothesis was inseparable from his radical political views. Before he became a professor of astronomy at the University of Glasgow, Nichol worked as a journalist writing on political economy for several liberal and radical magazines. His radicalism dominated his scientific and educational career.[84] In *Architecture*, Nichol clearly declared his enthusiasm for the science of progress, where he wrote: "Suppose we are yet mistaken; suppose the Nebular Hypothesis, with all its grasp, not to be the true key to the mystery of the origin and destinies of things, what is gained—what new possession—by that course of bold conjecture on which we have ventured to embark? This, at least, is established on grounds not to be removed. In the vast Heavens, as well as among phenomena around us, all things are in a state of change and PROGRESS."[85] In addition to being a popular author, Nichol was a successful lecturer. He propagated the nebular hypothesis not only in his writing but also in his lecturing. Nichol presented courses on astronomy at the Royal Manchester Institution several times; the syllabi of these courses in October 1850, September 1851, and April 1858 have been preserved in the Manchester Archives.[86] These syllabi clearly show that the science of progress, including the nebular hypothesis, was a significant subject in Nichol's lectures, such as in his presentation of topics titled "Speculations Concerning the Origin of the Solar System" and "Relation of Astronomical with Geological Epochs—Sketches of the Evolution of the Earth." Nichol's advocacy of the nebular hypothesis drew lecturing invitations from scientific institutions in the metropolis. For example, an advertisement stated that the Whittington Club and Metropolitan Athenaeum "have much pleasure in announcing that J. P. NICHOL ... has acceded to their request to deliver an Illustrative Course of Six Lectures ON THE PHYSICAL STRUCTURE OF THE SOLAR SYSTEM, during a short business visit to London."[87] Nichol's influence even spread across the Atlantic: he was on the lecture circuit in the United States between 1847 and 1848. Edgar Allan Poe was among the large audiences for Nichol's lectures in New York.[88]

Despite fervent interest or concern about the nebular hypothesis among scientific elites, the subject was absent in the curriculum of many popular lectures except for Nichol's. The nebular hypothesis never appeared, or at least not noticeably, in the syllabi of D. F. Walker, C. H. Adams, or John Wallis. These lecturers were content to show the current state of the solar system, rather than its possible evolutionary past. Perhaps this was due to uncertainty and their wish to avoid the controversy generated by the nebular hypothesis. These popular lectures, many of which were presented during seasons of Lent or Christmas, had a religious undertone. For a popular lecture aimed at a broad audience including children and parents, controversial issues that might involve materialist discourse like the nebular hypothesis were inappropriate and unnecessary. Among the many subjects demonstrating the achievement of human rational exploration of the world, Newton's cosmology is the pinnacle. Why bother to touch an uncertain hot potato? Besides, many astronomical showmen relied on authoritative publications as the source of their lectures. They were more familiar with conventional literature such as Ferguson's *Astronomy* than with controversial new accounts that lacked mainstream consensus.

John Herschel's comments on the nebular hypothesis represented the reserved attitude of establishment astronomy toward this controversial theory. As the son of William Herschel, John Herschel enjoyed prestigious status in British astronomy and was the leading man of science in early Victorian Britain. The rise of the nebular hypothesis, ironically, to some extent benefited from the advance of telescopic technology and the newborn discipline of nebular astronomy to which William Herschel had contributed. John wanted to defend his father's legacy and thus distinguished Herschel senior's cosmology from the inflated nebular hypothesis at that time. In the presidential address to the BAAS at Cambridge in 1845, John Herschel spoke cautiously about the nebular hypothesis.[89] On the one hand, Herschel acknowledged Laplace's idea that "it is impossible to deny the ingenuity" of the hypothesis, but on the other hand he pointed out that the theory still lacked acceptable evidence from observation. "If, therefore, we go on to push its application to that extent, we clearly theorize in advance of all inductive observation." He asserted:

> I am by no means disposed to quarrel with the nebulous hypothesis even in this form, as a matter of pure speculation, and without any reference to final causes; but if it is to be regarded as a demonstrated truth, or as receiving the smallest support from any observed numerical relations which actually hold good among the elements of the planetary orbits, I beg leave to demur. Assuredly it receives no support from observation of the effects of sidereal aggregation, as exemplified in the formation of globular and elliptic clusters, supposing *them* to have resulted from such aggregation.[90]

Because of this cautious view, Herschel did not like Nichol's *Architecture* and privately condemned it as a sensational potboiler. Similar criticism was presented by John Wallis. Although Wallis had not referred to the nebular hypothesis in his lecture syllabi, he wrote a pamphlet to attack it the same year that Herschel delivered the BAAS presidential address at Cambridge. Both Wallis's and Herschel's criticisms were in response to the sensationally radical view of the cosmogony, especially those infamous bestsellers such as *Architecture* and *Vestiges*, which propagated the nebular hypothesis and increased its popularity.[91]

Compared to the troubles caused by the nebular hypothesis, another controversial issue—the plurality of worlds and the implication of extraterrestrial life—was less problematic. Debates on the existence of extraterrestrial life are strongly associated with the assumption of terrestrial worlds outside the Earth because these hypothetical worlds are essential to providing the grounds for extraterrestrial beings, especially intelligent humanoids, to stand on. Neither of the two was a fresh idea. Similar arguments had been made since the ancient Greeks and had shown up in different eras and regions. For example, in Europe before the eighteenth century, Johannes Kepler, Christiaan

Huygens, and Bernard Le Bovier de Fontenelle had elaborated on the notion in their works. Copernicus's heliocentric cosmology, along with many later astronomical discoveries, were important factors that increased corresponding enthusiasm for a fully populated universe. By the mid-eighteenth century, the public had essentially accepted heliocentrism, and many intellectuals had adopted the idea of a plurality of inhabited worlds.[92] The English poet Edward Young's *Night-Thoughts* illustrated the influence of the plurality of worlds theory. The lines "Tell me, ye stars! ye planets! tell me, all / Ye starr'd, and planeted, inhabitants! what is it?" reflected this very idea.[93] In early nineteenth-century Britain, the eminent Scottish clergyman Thomas Chalmers was a significant proponent of inhabited extraterrestrial worlds. Chalmers presented a series of electrifying sermons on astronomy in Glasgow in 1815, and one of his topics was the plurality of worlds. His subsequent book based on the sermon, *A Series of Discourses on the Christian Revelation, Viewed in Connection with the Modern Astronomy* (1817), became an instant bestseller, reaching a tenth edition within five years. It effectively propagated the idea of inhabited extraterrestrial worlds.[94]

Although the idea of an inhabited extraterrestrial world could deprive humans of their unique status and thus threaten Christian beliefs, it did not disturb the religious communities as much as the problems caused by the science of progress. There was no consensus on this matter among Christian denominations, as they faced a dilemma. On the one hand, to accept the plurality idea would hinder the argument of humans as God's unique and favorite creation; on the other hand, to reject the plurality notion would suggest an unacceptable waste of God's power and creative abilities. In Scotland, the Presbyterian Church tended to embrace the idea of a plurality of inhabited worlds, since this notion could underline God's omnipotence as well as humans' insignificance. It stressed salvation through God's unrequited grace, and wonder that God would care for such insignificant beings as we humans.[95] As a Presbyterian minister and the leader of the later Free Church of Scotland, Chalmers and his *Discourses* represented this approach to the issue.

In contrast, Anglican attitudes toward the plurality of worlds were not so certain. The most influential opponent of the plurality notion was William Whewell. Initially, Whewell was open to the plurality concept and agreed with such a possibility in his Bridgewater Treatise in 1833. However, his position changed drastically afterward. He later wrote a volume titled *Of the Plurality of Worlds: An Essay* (1853) to refute plurality theorists including Chalmers. He severely disputed the plurality idea and the existence of extraterrestrial intelligence by making scientific and theological arguments.[96] Whewell's essay attracted widespread criticism, of which the strongest defense of the plurality idea was from David Brewster, a devout Presbyterian. The fierce debate between Whewell and Brewster drew more attention to the controversy surrounding Whewell's essay, and by 1859 more than twenty books and at least fifty review articles had responded to it.[97] Criticism of Whewell also came from within the Anglican community. For instance, Baden Powell, whose position was more liberal, disagreed with Whewell's extremist defense of the uniqueness of humans. Michael Crowe's quantitative assessment of the Whewell-Brewster debate shows that two-thirds of published works supported Brewster against Whewell.[98] This result indicates that the plurality idea, despite being contested, still enjoyed the majority of the public's support during this period.

Most popular astronomy lectures did not reflect the heated debates of the plurality idea among the scientific elites. Sentiments about the plurality of worlds were evidently prevalent in popular lectures and most lecturers showed approval. Unlike discussions of the nebular hypothesis, narratives of inhabited extraterrestrial worlds were common in the syllabi. This convention was, unsurprisingly, influenced by renowned authors including Chalmers and Ferguson, whose works expressed support for the plurality of worlds. Ferguson was especially crucial, since his works directly influenced the pioneers of stage astronomy including William Walker in the late eighteenth century before Chalmers preached sensationally in Glasgow in 1815. William Walker incorporated the plurality subject into his eidouranion lectures. Walker began his discourse with a passionate statement, "When we launch an idea into infinite space, and contemplate the systems without number that fill it, here indeed we have a subject truly worthy of the DEITY!" He then deduced the immensity of the universe from the vast quantity of stars, finally reaching the argument:

> The Sun's light could not therefore reach the fixed stars, and be reflected back again with their lustre; of course, then they shine by their own light;

if so, they shine as our Sun, and consequently are Suns themselves. Now as a principle of uniformity runs through the variety of nature, it is reasonable to conclude these Suns to be centres of [a] system like ours; and destined for the same noble purpose, viz. that of giving light, heat, and vegetation, to various worlds that revolve round them, but which are too remote for discovery, even with our best telescope![99]

Walker elaborated this argument just before the fifth scene of his program, in which the construction of the universe was shown: "the stars, disposed in constellations, and surrounded by concentric circles." He praised this idea as "infinitely too large for the human mind; or indeed for that of any created being!" The same sentiment appeared in Arnold and Bartley's lecture syllabus, in which the playwright quoted Ferguson's words as the conclusion: "Thousands of thousands of Suns, multiplied without end, and ranged all around us, at immense distances from each other, attended by ten thousand times ten thousand Worlds."[100] This very passage had already been repeatedly quoted, or copied, in many other popular astronomical publications between the late eighteenth and early nineteenth centuries. This phenomenon just shows that Ferguson's influence on popular astronomy prevailed for generations.

A later example of the plurality topic in popular astronomy lectures in the Victorian era was Ebenezer Henderson's *A Treatise on Astronomy* (1843). A son of a clockmaker, Henderson moved to astronomical lecturing from his original job as an artisan as his hero Ferguson had done. Henderson collected biographical materials and when he retired, he wrote a biography of Ferguson.[101] *A Treatise on Astronomy* was based on a course of twelve lectures presented by Henderson in London in 1835.[102] In his book, the chapter "On a Plurality of Worlds" focused particularly on the extraterrestrial life issue. Henderson summarized the arguments presented by many of his predecessors, making this chapter a concise and neat introduction to promoting the plurality idea.[103] He began his reasoning strategy with the nature of stars, and then turned to the other planets to explain that the moons of Jupiter and Saturn "can be of no use to the inhabitants of our earth." Next, he connected the analogy of satellites to the nature of stars, and asked: "Of what use to the Earth are those unseen colours, those periodical changes, those rotations? It would be presumptuous to imagine that such were ushered into existence merely for service to our Earth.... They must, therefore, have been created for a higher, for a far nobler purpose than for the use of the Earth." After all the necessary astronomical facts and analogy were deployed, he argued, the conclusion was clear:

> The great probability is, that every star is a SUN far surpassing *ours* in magnitude and splendour; they all shine by their own *native light*; they do not borrow their light from any body whatever. What a most powerful Sun that apparently little star Vega must be, when it is 53,977 times larger than our Sun! Our Sun, if removed to the distance of about two billions of miles, would appear far less in magnitude than the star Vega. The stars *being thus supposed to be* SUNS, *it is extremely probable that they are the centres* of OTHER SYSTEMS OF WORLDS, round which may revolve a numerous retinue of planets and satellites. Therefore there must be a *plurality of Suns*—A PLURALITY OF WORLDS.[104]

To strengthen the plurality argument, Henderson even prepared biblical fortifications for criticisms of materialism from religious people. He admitted that the Scripture is silent on a plurality of worlds, but it is "not at variance with such a supposition." He claimed, "Several remarkable passages which, when brought in connection with this subject, explain themselves with greater power of meaning." After discussing a few biblical passages, he concluded: "Reason, analogy, and the Scriptures furnish sufficient evidence to conclude that there is a plurality of worlds, and that they are inhabited by beings capable of appreciating the goodness, and *adoring* the wisdom, of the Creator."[105]

Henderson's discussion is an excellent account of the argument of cosmic pluralism in early Victorian Britain, which not only summarized many of his predecessors' arguments on the concept of pluralism but also applied the familiar rhetorical skills of natural theology such as sublimity, novelty, and analogy. His syllabus as well as Walker's and Bartley's all demonstrate some of the most absorbing characteristics of popular astronomy lectures at the time, which were often the key to commercial success and critical acclaim.

The curricula of popular astronomy lectures in the Regency and early Victorian eras were characterized by a mixture of convention and novelty. Conventional subjects related to Newtonian science in the eighteenth-century discourses on natural philosophy prevailed at all levels and in various types of astronomical lectures. This fact reflected the prestigious status of Isaac Newton as a national scientific hero in Britain. Astronomical events, such as comets and eclipses, were also captivating topics, especially for attracting audiences. Expositions of these astronomical phenomena were often accompanied by current events, previewing to and encouraging the audience to participate in astronomical observations, and were combined with reviews of similar historical events to illustrate the advance of human understanding of the universe. New discoveries and the latest scientific progress in the discipline were recurring subjects, too. Whether C. H. Adams's special coverage of the new planet, or the Royal Institution's discourses on a recent experiment or expedition, all represented attempts to incorporate scientific news into the lecture.

In addition to scientific knowledge of the physical universe, Christianity was a commonly explicit theme in the mainstream popular works of astronomy. The rhetoric of and issues related to natural theology strongly influenced contemporary popular astronomy lectures. Astronomy, long recognized as the sublime science, was deemed a suitable subject for demonstrating the magnificence of God's handiwork, and hence inspiring reverence for God. For their own reasons, connections existed between astronomy and religion had their own reasons: in addition to heavenly phenomena and the order of the universe itself as providing rich resources for such associations, the political and social milieu after the French Revolution also prompted the establishment to promote a "safe" science to combat subversive radicalism. The controversies surrounding the nebular hypothesis and the plurality of worlds, the two most debated issues in contemporary astronomy, reflected such concerns.

We have seen which subjects appeared in popular astronomy lectures. In chapter 5, we explore how lecturers conveyed knowledge and rendered sublimity to the audience by relying on their appropriate use of apparatuses (i.e., visual aids) in their lectures.

CHAPTER 5
APPARATUS

The scientific celebrity Carl Sagan, renowned for his popular works *Cosmos* (1980) and *Pale Blue Dot* (1994), was invited to give Christmas Lectures at the Royal Institution in 1977, before he wrote the two books. The title of Sagan's Christmas Lecture series was "The Planets," which narrated stories about the solar system from the Earth to Mars and planetary systems beyond the Sun.[1] Later, Sagan must have developed the content of his Christmas Lecture series into his two bestsellers. He was famous for breaking the norms of academia and his well-publicized scientific advocacy in the mass media. Despite his innovative traits, Sagan began his Christmas Lecture in a rather solemn, old-school fashion. From its rhetoric to its structure, the lecture conveyed an implicit sense of inheritance and tradition, as if it were nothing fresh, but a continuation of astronomical lectures from the past century.

Many subtleties in Sagan's Christmas Lecture symbolized the sense of continuity with the tradition of scientific lectures. For instance, in his opening remark, Sagan paid tribute to Michael Faraday as this was the 150th anniversary of Faraday's first Christmas Lecture. The apparatus that Sagan demonstrated at the beginning—an orrery—was also a gesture toward tradition. Demonstrations of orreries had been a classic act in popular astronomy lecturing for more than two hundred years since the device was invented in the early eighteenth century. Although we do not know whether Faraday ever used an orrery in his lectures, orreries were still commonly used by popularizers of astronomy in Faraday's time in the mid-nineteenth century. In addition to the orrery, Sagan also used other visual aids, including slide shows. Like its predecessor, the magic lantern in the nineteenth century, a slide projector commonly presented slide shows in Sagan's time, but has become an antique for today's younger generations. Electronic video display devices and computer software such as Microsoft PowerPoint that have replaced slide projectors can be counted as the apparatus's distant cousin. The technology is different, but the functions are similar.

From the nineteenth century to the present day, popular astronomy lecturing has been characterized by continuity in the way it has been communicated to the public. Astronomy has been a visual science with a rich history and practice. This is not only because of the astounding view of dazzling night sky or closeups of celestial objects but also because the language behind scientific research on astronomy—mathematics—is often quite abstruse to most people. For centuries, astronomy lecturers have relied on visual aids to communicate knowledge, whether in geometrical or descriptive ways. In addition to conveying knowledge to give a rational understanding of the universe, astronomical visual aids must also evoke a sense of wonder. They are powerful tools of the sublime science as they provide sources of beauty and imagination directly through visual experience. As Jan Golinski indicates, the technical methods of the Walkers' eidouranion—the prototype of the theatricalized astronomy lecture—created overall aesthetic effects and allowed spectators not only to witness the sublime but also to feel it.[2]

This chapter examines the development of the visual aids used by astronomical lecturers from the time of James Ferguson in the mid-eighteenth century to the time of Michael Faraday in the mid-nineteenth century. The two major branches of astronomical visual aids were mechanical devices primarily driven by gears, such as the mechanical orrery, and optical devices based on optical principles and effects, such as the magic lantern and lantern slides. The importance of these apparatuses in the history of popular astronomy as well as the ways in which popularizers used the visual aids to achieve ideally lucid and sensational effects exemplify continuity in popular astronomy at the outset. Nevertheless, the notion of continuity does not mean the devices themselves have not changed much over the period. They evolved, derived products of different types and functions, and even produced hybrids. These changing versions of visual aids over time demonstrate that mechanical innovation was always of central importance in attracting audiences through novel and spectacular devices.

THE INSTRUMENT OF POLITE SCIENCE PAR EXCELLENCE

Orreries were a major branch of visual aids used by astronomy lecturers in the eighteenth and nineteenth centuries. The mechanism reached iconic status as "the instrument of polite science *par excellence*."[3] An orrery is a mechanical model of the solar system. A prototype design was invented by the English clockmaker George Graham, with the assistance of his mentor Thomas Tompion, circa 1710, although similar devices showing the planetary system had been independently contrived by the Dutch polymath Christiaan Huygens and the Danish astronomer Ole Roemer before this date.[4] If we revisit ancient times and broaden the definitions, we see that humans in different civilizations long had aspirations to use machines to measure or model the heavens' movements. The Antikythera mechanism from ancient Greece in about the second century BC, astrolabes from Islamic Golden Age, the Chinese water-driven astronomical clock tower by Su Song circa 1092, and astronomical clocks in medieval European cities and cathedrals, all belong to this umbrella category that Henry C. King describes as the family of "geared astronomical mechanisms."[5] To clarify and focus our discussions, here I confine the scope of interest to "modern" orreries, that is, those devices derived from George Graham's prototype model since the early eighteenth century.[6]

Graham later presented his prototype to the celebrated instrument maker John Rowley. Inspired by Graham's design, Rowley improved it and made his own device, which he then presented to his patron Charles Boyle, the 4th Earl of Orrery, around 1712. Since then, this novel machine has been named after the Earl in the English language.[7] The earliest orreries made by Graham or Rowley, however, were not like the familiar one that simulated the entire planetary system. They showed only the Sun, Earth, and Moon. Later, the alternative term "tellurian" (also spelled "tellurion" or "tellurium") was often used to describe the device showing only the Sun, Earth, and Moon, particularly in comparison to the "planetarium," which included other planets, as Benjamin Martin's design shows (see ch. 1, fig. 1.3).

Rowley and his assistant Thomas Wright made significant modifications to their models. Between 1715 and 1728, they added attachments of other planets and moons onto the device, as the complexity of clockworks increased. The resulting product was the more elaborate "grand orrery," a classic configuration that had also been described in the works of philosophical lecturers such as John T. Desaguliers and Benjamin Martin.[8] The grand orrery became an iconic tool of Enlightenment science as the painting by Joseph Wright shows (see ch. 1, fig. 1.1). A surviving example of a grand orrery, made by George Adams (senior) circa 1750, is on display at the Whipple Museum in Cambridge, England.[9] This grand orrery is 1.05 meters in diameter. It has a twelve-sided wooden base made of mahogany or ebonized oak; each side of the base has a panel with hand-illustrated gilded signs of the zodiac. The twelve ball feet, the discs carrying planets, and the circles of the armillary hemisphere are made of brass; and the Sun, Earth, and other planets are made of brass or ivory. Overall, this grand orrery is a combination of craftsmanship and fine arts: it consists of expensive materials, has a solid and heavy structure, but it is elegantly decorated.

Around this time (the late 1710s), philosophical lecturers began using orreries as educational aids. As early as 1717, Desaguliers had described Graham's "Planetary Machine" in his lecture treatise. Other contemporaries also praised the instructive value of orreries. For example, William Deane remarked on the educational function of the grand orrery: "THOSE Gentlemen and Ladies who delight in the Study of *Astronomy* and *Geography*, will, by seeing this *Grand Machine*, comprehend at one View the Reason of the several *phenomena*, or Appearances, in the Heavens, resulting from the various Motions of the Bodies which compose this *Solar System*; and will edify more from a few Lectures, than by a Year's close Application to Study."[10] However, orreries were expensive. A grand orrery could cost hundreds of guineas. According to a 1772 catalog of instruments produced under the inspection and direction of George Adams, the price of a large orrery was in the range of 130 guineas to £1,500 (about 1,428 guineas), depending on the demands and wishes of the customer.[11] A smaller planetarium and a tellurian, which was not as elaborate and multifunctional as the grand orrery, still cost £18/18/0 and £26/15/0, respectively.[12] The extraordinary level of the highest price might be somewhat exaggerated, as George Adams was the instrument maker to King George III and was expected to provide high-end service. Nonetheless, it could still reflect the status of the grand orrery as an extremely luxurious commodity. The historian of scientific instruments John Millburn summarizes that the prices of orreries (including various types from the smallest to largest) from several instrument makers in London during the period were in the range between £25 and £250.[13] Such high prices made most orreries affordable only to institutions or wealthy private customers. Recall that the value of 1 guinea in the 1770s was equivalent to the weekly wage of a journeyman silversmith, or the cost of twelve bottles of fine Portuguese wine.[14] Before the late eighteenth century, the orrery remained largely a luxury curiosity through which artisans showcased their craft, although it had been widely adopted as a necessary apparatus in public philosophy lectures.

Many instrument makers and philosophical lecturers attempted their own modifications and improvements to the orrery. Some tried to make the apparatus simpler and less expensive, rather than more complex, sophisticated, and costly. The most influential products in this direction (see chapter 1), were manufactured by Benjamin Martin. Martin asserted that because of the very expensive prices of orreries, "this most useful machine is not so common as one might wish it were." It was a shame because the orrery could greatly enhance the ability to convey "an easy and adequate idea" of the world's system and principles of astronomy and geography, which was an essential part of the education that "our English youth should have."[15] As early as 1747, Martin had contrived a "double-cone" wheelwork mechanism and described it in the textbook *Philosophia Britannica*, but he did not make orreries containing this mechanism until he opened a shop trading as an instrument maker on Fleet Street in London in 1756. He further described the new design in detail in *The Young Gentleman and Lady's Philosophy* (1759), as figure 5.1 shows, and in *The Description and Use of Both the Globes, Armillary Sphere and Orrery* (1762). The mechanism was a set of separate coaxial tubes surrounding a vertical Sun column, each tube having an arm carrying a planet ball at its upper end and a gear wheel (planet gear) fixed to its lower end. Each planet gear meshed with a corresponding coaxial driver gear on a parallel axle. Thus, the planet gears formed an inverted cone (Saturn's gear, the largest, was at the top), while the driver gears formed an upright cone.[16] Martin believed that this concise structure could make it possible to produce a portable planetarium at a small cost. An extant example of Martin's double-cone orrery, made around the mid-eighteenth century, is pre-

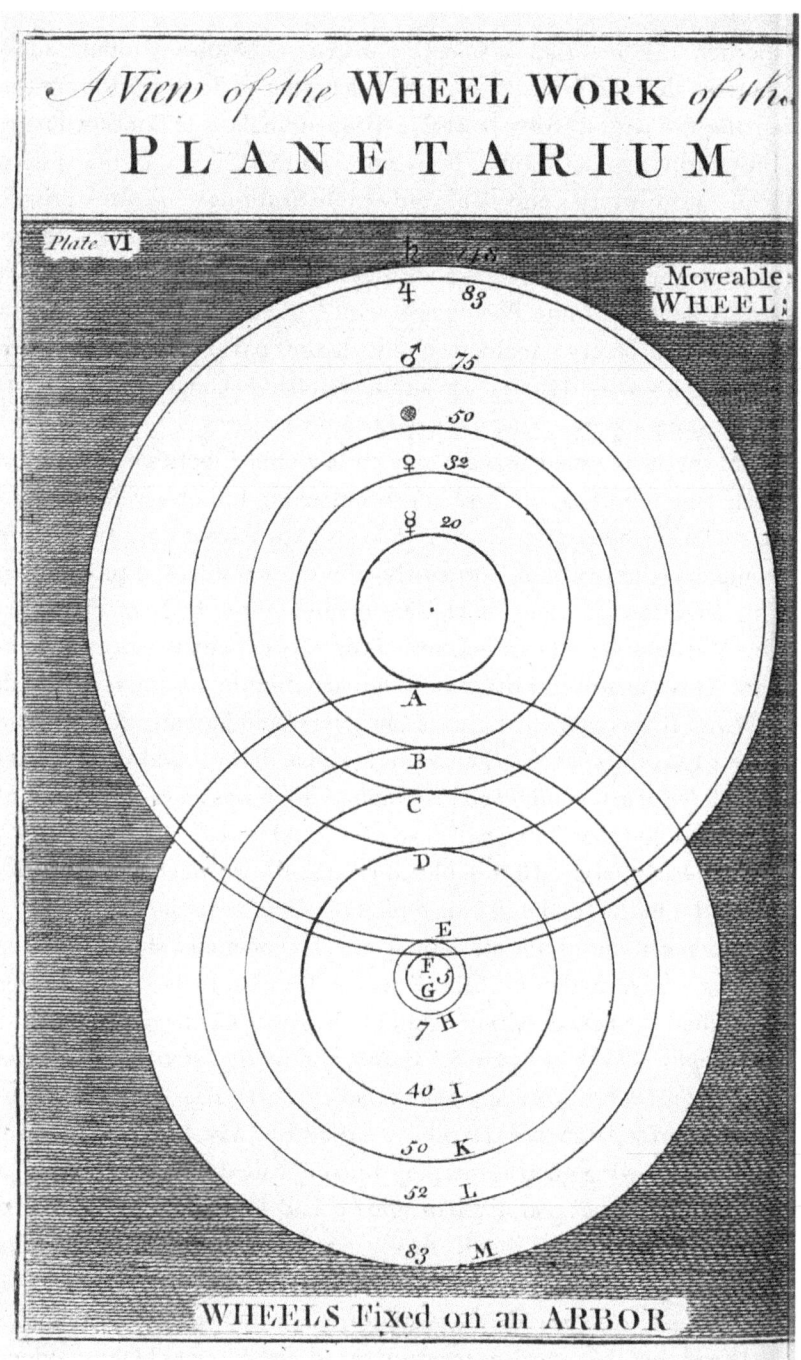

FIGURE 5.1. Benjamin Martin's double-cone wheelwork mechanism. Plate from *The Young Gentleman and Lady's Philosophy*, by Benjamin Martin, London, 1759. Image digitized by the Internet Archive, with funding from the Wellcome Library.

served in the collections of the Science Museum in London (inventory no. 1919-463).[17]

This type of double-cone mechanism was not original to Martin, however. The same configuration had been employed by Roemer over half a century earlier, and the details of Roemer's planetariums were illustrated and published in 1735. Nevertheless, since the accounts were published in France and Denmark, Martin was probably not aware of Roemer's planetariums. One subtle difference between Roemer's and Martin's plans was in the gear ratios; this made the orbital periods of planets in Roemer's model closer to the real periods. Although Martin's model was less accurate, its gear trains would have been more easily implemented.[18]

Martin's efforts to simplify orreries did effectively reduce their cost. His catalogs published soon after the Fleet Street shop opened in 1756 showed that the price of the cheapest manual orrery was £2/12/0, and the prices of planetariums ranged from £5/5/0 to £22/1/0. The cheapest model was very simple and had no wheelwork at all.[19] He continually made modifications for orreries, adding attachments and improving accuracy based on the double-cone mechanism. In his last tract on the orrery, *The Description and Use of an Orrery of a New Construction* (1771), Martin proposed a "modular" design to divide the orrery into several components, including the planetarium, tellurian, lunarium, Jovian system, and Saturnian system (see ch. 1, fig. 1.3). Customers could purchase different components according to their needs. The basic set of the planetarium cost £12/12/0; the two-piece set of the planetarium and tellurian cost £21; the three-piece set of the planetarium, tellurian, and lunarium, cost £35; and a complete set of all the components cost £157/10/0.[20] He claimed the expense of such a modular construction was far less than that of a conventional grand orrery, but could demonstrate far more celestial and terrestrial phenomena. Perhaps to counter doubts about the accuracy of his simplified models, he also penned a detailed piece on the theory and calculations for the wheelwork of an orrery in this tract. Martin's support for simplified orreries did not necessarily mean that he gave up making grand orreries. In 1764, he was commissioned by Harvard College to make a grand orrery, and the product was shipped in 1767. It was a more elaborate model akin to the classic design of Rowley and Wright, rather than a simpler double-cone planetarium. The price was nearly £90, still cheaper than the usual cost of the same type by contemporary London instrument makers. This grand orrery by Benjamin Martin is extant and on display at the Putnam Gallery at Harvard University in Cambridge, Massachusetts.[21]

Although Martin's design of a modular planetarium with separate lunarium and tellurian attachments was promising, John Millburn indicates that the most popular standardized model on the market after the 1780s was a hybrid "Jones-type" design, rather than Martin's type. In this hybrid design, the Earth, Moon, and inferior planets were operated by exposed wheelwork mounted on a combined tellurian and lunarium arm, and only the superior planets were driven by double-cone wheelwork.[22] The hybrid "new portable orrery" was invented by William Jones around the early 1780s. Jones constructed the wheelwork mechanism of the tellurian following the principle contrived by Ferguson about four decades earlier, but the idea of separate tellurian and planetarium attachments was apparently inspired by Martin's more recent design.[23] William Jones also acknowledged Martin's contribution to the modular configuration of the orrery, as he attributed such a complete set to be "contrived by the late Mr. B. Martin" in his early catalogs.[24] This very type of portable orrery with a hybrid structure was manufactured and sold by many commercial instrument makers in the late eighteenth and nineteenth centuries, such as W. & S. Jones. The joint business of William and Samuel Jones was one of the more prominent and enduring instrument-making firms in the first half of the nineteenth century. William Jones, the elder sibling of the Jones brothers, had manufactured and traded scientific instruments in Holborn, London, since the early 1780s, before the younger brother Samuel joined the venture and formed the W. & S. Jones company. Even after William's passing in 1831, there were still products branded as W. & S. Jones until 1859.[25]

An example of this hybrid type of portable orrery made by William Jones, produced circa 1794, is preserved in the National Maritime Museum in London (fig. 5.2).[26] It has two separate wooden baseboards for the tellurian and planetarium attachments, and the planetarium is very simple without wheelwork. William Jones cut the prices even more, as the tellurian part cost only £1/8/0, and with the planetarium part the price was £3/10/0. The prices of a complete set with both attachments, packed in boxes, ranged from £3/13/6 to £6/6/0 accord-

FIGURE 5.2. Portable orrery with tellurian attachment, made by William Jones, circa 1794. National Maritime Museum, inventory no. AST1062. © National Maritime Museum, Greenwich, London.

ing to the different sizes and wheelwork.[27] Like most commercial instrument makers, W. & S. Jones also made and sold various types of orreries, from the simplest portable model to the classic grand orrery. One of the larger orreries made by W. & S. Jones after 1789, which is also held in the National Maritime Museum, is a fine example of Martin's three-piece modular orrery (figs. 5.3, 5.4, and 5.5).[28] This stylish orrery consists of three separate attachments and is identical to the design proposed by Martin in 1771. It also matches William Jones's description of "a complete planetarium, tellurian, and lunarium, all elegantly in brass, shewing the motions completely by wheelwork... the earth a 3 inch globe" in the catalog, at a price of £37/16/0.[29] The three attachments can be mounted on a circular drum-like base standing on a column and three legs. The brass base is about 30 centimeters high and the overall height is nearly 50 centimeters with the tellurian installed. This height allows the instrument to be seen from a distance by spectators, making it ideal for demonstrations before a larger au-

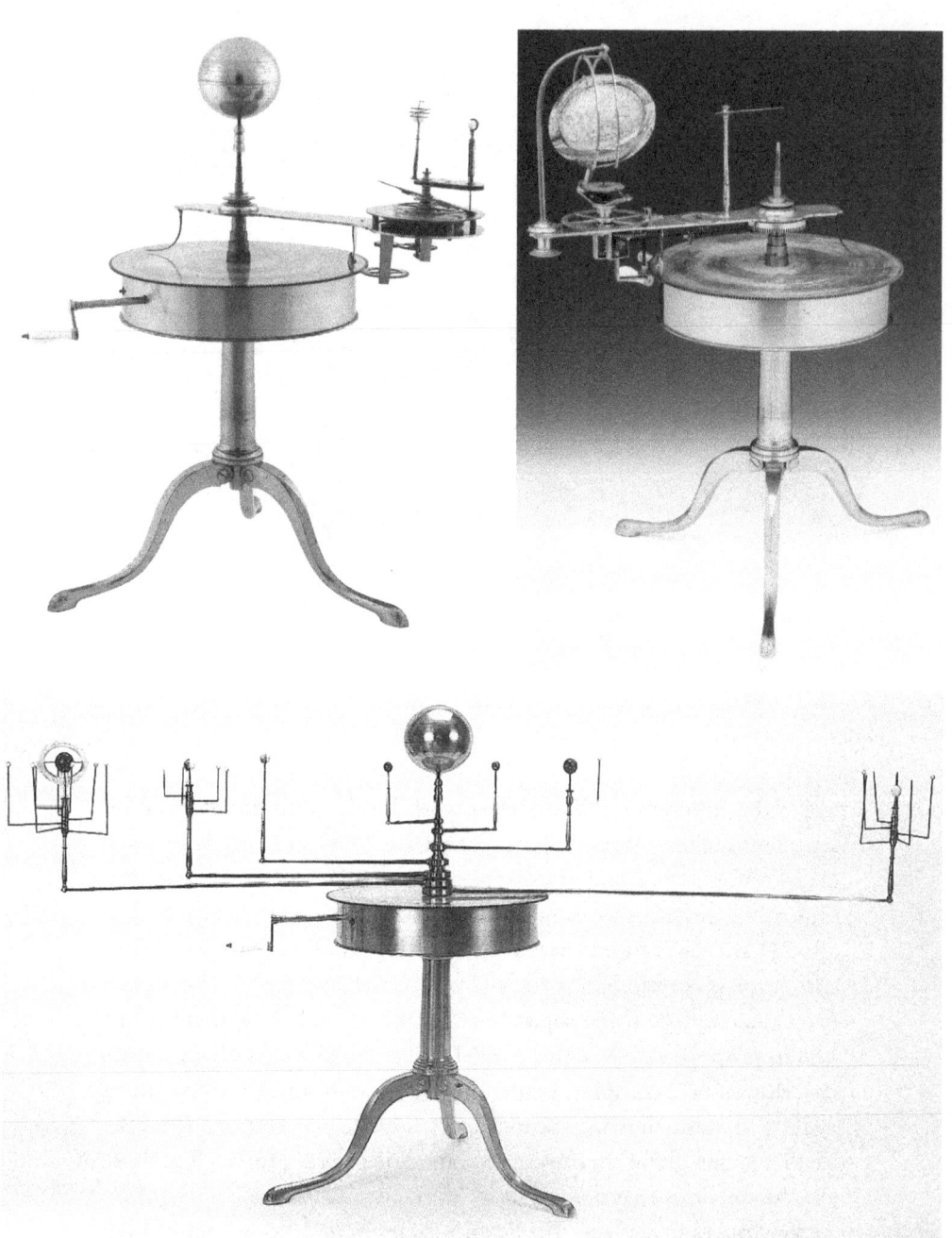

FIGURES 5.3, 5.4, AND 5.5. Three-piece modular orrery made by W. & S. Jones after 1789. (5.3) Lunarium; (5.4) Tellurium; (5.5) Planetarium. National Maritime Museum, inventory no. AST1055. © National Maritime Museum, Greenwich, London.

dience. In comparison, the smaller and cheaper Jones-type portable orrery was more suitable for use in a domestic environment as parlor entertainment or for a private tutorial.

Martin's design of separate attachments showing the systems of Jupiter or Saturn also inspired other orrery makers. An intriguingly unique artifact, also held in the storage collection of the National Maritime Museum (AST1066), is an orrery specifically focusing on Saturn and its moons. This orrery, with the engraving "W. LACY Invenit & Delineavit 1779" on its central brass plate, was the craftwork of William Lacy, who was then trading as a philosophical instrument maker at Drury Lane in London.[30] It is not an easily portable gadget: the wooden circular base, supported by three legs, is 35 centimeters in diameter and 25 centimeters high; the total length including the longest arm is 53 centimeters. Although this orrery also included the Earth, Moon, and Jupiter's system, its most impressive feature was Saturn's system. It had a large, prominent ivory ball, to which attached a brass flat ring, representing Saturn and its rings. There were originally five small ivory balls, representing the moons of Saturn known at the time when the orrery was made (1779): Tethys, Dione, Rhea, Titan, and Iapetus, by the sequence from the inner to the outer part, though the Titan ball is missing. The arms carrying these moons were mounted on an exposed modified double-cone gearing mechanism, allowing the moons to rotate around Saturn. There is also a curious sign of tinkering: an extended brass plate is attached to the arm carrying Saturn's system, yet it is not connected with the gearing mechanism. Two additional moons, presumably Mimas and Enceladus, were attached on the plate, but one has been lost. Mimas and Enceladus were both discovered in 1789, so this extra attachment was likely added after that date.

This curious apparatus matches the instrument that Lacy introduced in his book *An Introduction to Astronomy* (1777), where he described his invention of lunariums that show "all the phenomena of Saturn's ring and five moons, as seen from the earth and sun, in every year." He made similar models representing Jupiter and its four moons. The price of each device was 5–10 guineas as listed in the catalog.[31] However, judging by the large size of the object in the museum and its sophisticated gearing mechanism, it was unlikely that unusual craftwork like this could be obtained for a lower cost. It is also unclear whether Lacy had sold any model of the same type. This unique object was perhaps something experimental or purposely built for a showcase. In any case, Lacy's device, along with other variations of Martin-type orreries, exemplified the trade's vigorous ingenuity toward the late eighteenth century.

Orreries in the Crystal Palace

After the booming development of orreries in the second half of the eighteenth century, what was the status of this iconic kind of astronomical visual aid in the nineteenth century? The Great Exhibition of 1851 is a good starting point for showcasing Victorian material culture and the significance of scientific instruments in contemporary education. The Great Exhibition was indeed a showcase—an unprecedented and sensational one—gathering all sorts of arts, crafts, and industrial manufacturing in the world inside one magnificent purpose-built glass house. It provided an extraordinary opportunity for a contemporary visitor to view the best of the industries in a single place.[32] The Great Exhibition also presented some of the finest examples of the apparatus used by astronomy popularizers at the time.

From May 1 to October 15, 1851, the Great Exhibition at Hyde Park made a memorable summer for Londoners. Over 4 million visitors, of which 58,427 were foreigners, arrived in Britain's capital during this period; these figures represented 50 percent and 276 percent increases, respectively, over the same period in the preceding year. The number of visitors to London was equivalent to as much as one-fifth of the population of Great Britain. Eventually the number of admissions to the exhibition during its run totaled more than 6 million, which included repeat visits by season ticket holders.[33] This was the first international fair to exhibit "the Works of Industry of All Nations," according to its official title, on such a large scale. Along with the United Kingdom and its overseas territories and colonies, more than 30 foreign states attended the Great Exhibition, having either manufacturers' delegates or local goods and raw materials to "represent" the nation. Some countries listed in the exposition's *Official Catalogue*, such as China, did not actually send official delegates to the exhibition. Chinese articles and teas imported by Western collectors or traders were nevertheless on display there. Another example, the Society Islands, a South Pacific archipelago that was then a French protectorate, had no official delegates and was represented by several pieces of bead-dresses and clothes made by Aboriginal women.

During the summer of the Great Exhibition, newspapers and magazines focused on the Crystal Palace, the magnificent glass and iron edifice housing the exhibition. The public fever persisted, and the "void" after the exhibition closed haunted Londoners. "At last the fact has become history," as one essay in the *Illustrated London News* described the exhibition, "the one great sight of London, of England, of Europe, towards which every eye was turned, and which formed the one monster topic of discussion and admiration, overshadowing and dwarfing all others." This great spectacle that existed and flourished "but a week ago . . . as it were, of our daily lives and thoughts and sensations," is now over and passed, becoming "a thing of memory, a vision to be mentally recalled." The journalist lamented: "The blank is curiously great."[34] London became the capital of all the world in that golden summer, and the Crystal Palace was a congress gathering the wealth of nations.

On the eve of the closure of the Great Exhibition, the London newspaper the *Standard* reviewed the exhibits and urged its readers to grab the last chance to witness this wonderful spectacle. The paper cited several passages from *The Lily and the Bee: An Apologue of the Crystal Palace of 1851* written by Samuel Warren to introduce the exhibits that were "so philosophically and poetically illustrated" in the book.[35] The *Standard* journalist praised the book's author for having "taken a view at once so bold, novel, and instructive" of the objects in the Crystal Palace. One of the leading objects Warren described was the Vertical Orrery on the southern side of the organ in the western nave. He used this orrery as a backdrop to tell an imaginary "ghost story." The *Standard* journalist apparently enjoyed this story and incorporated an extract from Warren's book into the review:

> Two children, says the author, were standing opposite this Orrery, in the daytime, "merrily telling each other how the planets went round the sun," and even their times and distances the urchins knew—but of the wasting thought, and which, of sleepless centuries, to tell them what they told so trippingly, "recked they nought." At midnight were seen standing before the Orrery, the "sorely amazed ghosts" of the ancient astronomers; seeing it subvert all their own systems—those "of Chaldean and Egyptian sage,

and Greek philosopher," and "melting their ancient wisdom into air." Presently Newton, a radiant spirit, is seen explaining to the ancient astronomers the prodigious discoveries of modern astronomy, at the same time paying a majestic homage to revelation.[36]

Warren imagined that when the Crystal Palace was empty at midnight, ghosts from the past appeared and visited the spectacular exhibition the way visitors did during the day. Among the ghosts were many great astronomers and philosophers, including Pythagoras, Ptolemy, Copernicus, Galileo, Kepler, and Descartes. These spirits gathered in front of the Vertical Orrery, and Sir Isaac Newton, the most prominent among them, showed the correct system of the universe to his fellow sages.

For contemporary readers who had been accustomed to the narrative of the history of astronomy and cosmology by authors like Ferguson and Walker since the eighteenth century, Warren's story about a march of intellectuals with Newton at the apex was certainly familiar. Moreover, Warren made vivid use of the rhetorical technique of contrasts: day and midnight in the same spot; the living and the ghosts; children and philosophers. On the one hand, the spread of modern astronomical knowledge was so effective that even the "urchins" knew the distances and orbits of planets. On the other hand, the correct system that the children babbled about was easily taken for granted, like Warren's lament that the children were not aware of the efforts and genius behind common sense. This passage also reflects the sublimity of the English national hero Isaac Newton and the religious meaning behind the cosmology he presented. By speaking the correct system aloud and paying homage to the Deity, Newton ought to shine and eclipse other sages.

Neither Warren nor the *Standard* journalist explicitly indicated the details of the Vertical Orrery they referred to. Another source, *Tallis's History and Description of the Crystal Palace* (1852), provides some clues. This extensively illustrated three-volume compendium was a definitive reference work of the Great Exhibition and its displayed objects. One chapter of the compendium, "Telescopes, Orreries, Globes, and Model Mapping: *From the Juries' Report*," describes the astronomical and geographical instruments in the exhibition by citing comments from the official juries' report.[37] The organizers of the Great Exhibition classified all the exhibiting articles into thirty categories called "classes." Orreries and other astronomical apparatuses belonged to the class of philosophical instruments. The jurors of this class included several prestigious British and foreign men of science such as David Brewster, John Herschel, the Swiss physicist Jean-Daniel Colladon, and the French astronomer Claude-Louis Mathieu.[38]

What comments did the jury make about the exhibits of the subclass "orreries, planetariums and astronomical machines?" Sadly, the jurors were not impressed by the general result. They lamented that the time and ingenuity devoted to the several machines of this subclass "had not been better directed" since the instruments on display did not indicate any improvement over the many that had been constructed. However, one exception in this subclass was Facey's Orrery, "a vertical orrery of large dimensions, made by a working man, after his own design." This vertical orrery was constructed by a Mr. Facey, who, "by becoming a member of [the] Temperance Society, felt it necessary to do something to fill up the vacancy of his idle hours." Accordingly, Mr. Facey turned to the study of astronomy, and he built the vertical orrery as the result of his labor and ingenuity.[39] *Tallis's History and Description of the Crystal Palace* describes Facey's Orrery:

> This ingenious piece of mechanism was designed to assist students of astronomy, and was nine feet in diameter. It represented the principal bodies in the solar system, and showed all the planets and other attendant satellites revolving round the sun in their proper order. To effect this in the machine, it was necessary to employ no fewer than 194 accurately adjusted wheels to other apparatus fitted up on a new principle. In the limited space within which the exemplifications were confined, it was of course, impossible to show either the comparative sizes or distances of the heavenly bodies. The orrery, however, gave a general idea of the relative positions and revolutions of the planets and satellites, whilst a gentleman attended and gave a description of some particulars relating to them.[40]

The jury of the Great Exhibition awarded a Prize Medal to Mr. Facey for "the ingenuity displayed by him in the construction of this orrery."[41] Furthermore, although the report might be slightly exaggerated, it is understood that Facey had never seen any kind of orrery before constructing his own. His ingenuity and talent therefore deserved the

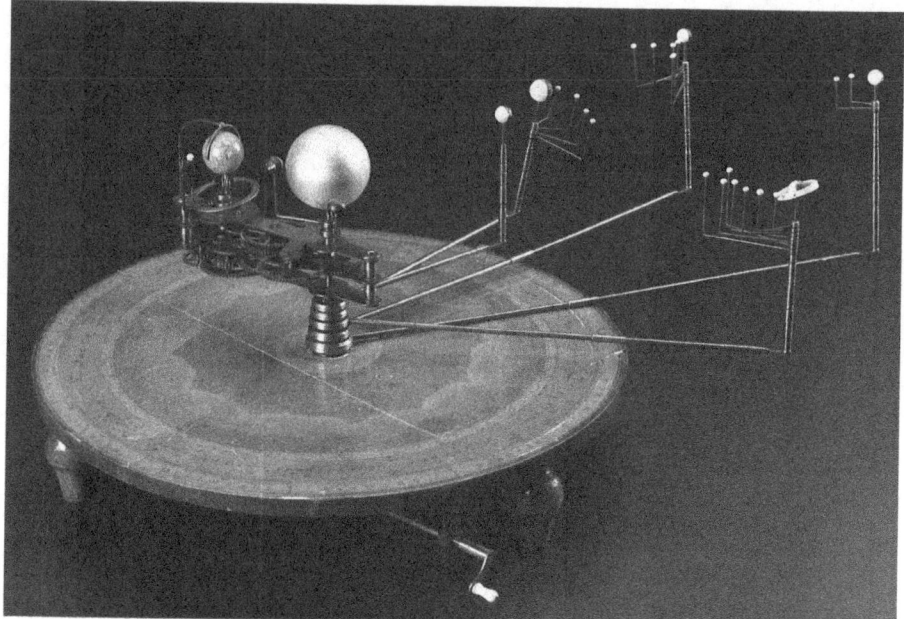

FIGURE 5.6. Portable orrery with brass gearwork and wooden base made by Newton & Son (Newton & Company after 1851), 3 Fleet Street, London, 1851–56. Science Museum, inventory no. 1869-48. © The Board of Trustees of the Science Museum, London.

award. The merit of Facey's Orrery, however, probably lay in its moral value rather than its technical quality. As many other nineteenth-century accounts like to explain, Facey's story was another perfect example of an artisan raised in humble origins who had managed through self-education to devote himself to a better, more useful life. Even better, Facey transformed himself from a drinker who was wasting his talent on bad hobbies to an industrious person acquiring beneficial knowledge. It suited the Victorian morals of personal improvement and social good. Facey's Orrery was the epitome of this didactic function served by astronomy.

Since Facey's Orrery was the only large-size vertical orrery registered in the Crystal Palace's official catalog, it was likely the apparatus that Warren and the *Standard* journalist referred to. Its size, 9 feet in diameter, made it suitable for display in a lecture room to a large group of children and novices. Nevertheless, Facey's Orrery was not the only exhibit with similar demonstration purposes. More than ten other orreries and planetariums were displayed in the Great Exhibition. Not as large as Facey's, most of them did not exceed 20 inches in diameter and were designed to be used on a table. For example, a manufacturer F. Plant of Nottingham exhibited an orrery with the Sun represented by a luminous body; the jury commented that this device would have a great advantage for use in a darkened room to demonstrate the seasons, phases of the Moon, and other natural occurrences. Newton & Son, a famous firm of globe makers based on Chancery Lane and Fleet Street in London, also exhibited a planetarium for educational purposes. Figure 5.6 depicts a portable orrery manufactured around the same time by the firm Newton & Company, which was presumably similar to the product exhibited by Newton & Son.[42] The simple design of this planetarium, which followed the layout of the Jones-type orrery, was common in the contemporary instrument trade. The juries' report remarked that the Newton & Son's planetarium was exhibited for "cheapness."[43] Some manufacturers were from overseas dependencies or foreign countries: Le Feuvre of Channel Islands presented an orrery designed for the use of schools; Massett of Switzerland exhibited a planetarium with extremely simple construction and remarkable cheapness, which received an Honorable Mention from the jury.[44]

The jury's generally low evaluation of the orreries in the Great Exhibition shows the stagnation of the orrery manufacturing industry and reflects the opinion gap between commercial manufacturers and scientific experts. Orrery making had been a mature trade since the late eighteenth century, but the layout of the products did not continue being refined. Most of the established London orrery manufacturers in the nineteenth century were content to adopt the basic design of what the historians Henry C. King and John Millburn call "Jones-type" models, which were derived from the portable orrery invented by Benjamin Martin in the mid-eighteenth century and eventually refined by other manufacturers, including George Adams and William Jones. Jones-type orreries were simple, affordable, and easy to manufacture. They had been popular among customers since their introduction in the early 1780s and were virtually unaltered until the 1850s.[45] As King and Millburn's survey of later orreries made by manufacturers other than W. & S. Jones shows, nineteenth-century commercial products had few additional improvements and the crafters had made only a few trivial changes such as adding newly discovered planets.

The conformity of orrery manufacturing in the nineteenth century indicates a drastic change in both the trade and the market. Before the late eighteenth century, artisans sought patronage from aristocrats or the wealthy population for constructing an exquisite clockwork machine. Clockmaking was a lengthy process and required extensive technical skill; therefore, the production was usually the way for crafters to show their artisanship. Yet the time of delicate handicrafts had passed. The industrial age was also the age of mass culture and mass manufacturing. The growing literate population from both the middle- and working classes now became greater part of customers for science. These new classes of patrons needed to learn essential scientific knowledge rather than collecting luxuries. For elementary teaching purposes, any overly ornamental or complex part of the apparatus was deemed unnecessary, but scientific accuracy could also be compromised due to simplification. Customers needed cheap and robust products, so manufacturers were content to copy a simple, affordable, standardized design that had been proved to be workable, like the Jones-type orreries.[46] The instrument-making trade was transitioning from custom-made handicrafts to mass-produced manufacturing. The orreries on display in the Crystal Palace therefore reflect the divergence between manufacturers, who stood for commercial profits, and expert judges, who promoted scientific values.

Although most orrery manufacturers in the early Victorian period were satisfied with a simpler, standardized design for merchandise, some instrument makers were still determined to improve the orrery by producing more sophisticated models that corresponded to real planetary motions. Unlike the commercial exhibits at the Crystal Palace, the orreries of William Pearson and John Fulton were two prominent early nineteenth-century examples of state-of-the-art artisanship and ingenuity displaying a more accurately scientific representation of the heavens.[47] They were, perhaps, the kind of models that the juries of the Great Exhibition wished to see.

The first of the two, William Pearson (1767–1847), was an instrument maker and astronomer with a close affiliation to several learned societies and scientific institutions. In his early career, Pearson was a schoolmaster at a grammar school in Lincoln and later moved to London for posts at other preparatory schools and businesses. He also became a clergyman and a local magistrate. Apart from his school and parish duties, he became famous in scientific circles for his mechan-

FIGURE 5.7. Group portrait of William Pearson and his wife and daughter, by Thomas Philips, after 1800. By permission of the Royal Astronomical Society, London.

ical expertise in scientific instruments and horology.[48] He began his scientific pursuits by constructing astronomical instruments such as astronomical clocks and orreries. He was one of the original proprietors of the Royal Institution and built several planetary machines for demonstration in a large lecturing space. The Royal Institution's earliest natural philosophy lectures by Thomas Young had used Pearson's apparatus. He was elected a fellow of the Royal Society in 1819. Later, he contributed to the founding of the Royal Astronomical Society, drew up the society's rules, and served as the treasurer during its first decade. A group portrait of Pearson with his wife and daughter (fig. 5.7), in which he proudly points at an orrery of his design, is hung on the wall of the fellows' common room at the present-day site of the Royal Astronomical Society at Burlington House. As an avid orrery maker affiliated with metropolitan scientific elites, Pearson publicly criticized the flaws of scientific inaccuracy in general contemporary orreries, as his articles on planetariums and orreries in the *Rees's Cyclopaedia* show. To remedy the flaws, he built several improved devices, which were notable for their sophistication and accuracy.

Two representative orreries that Pearson designed were "Orrery for equated Motions in three Parts" and an "improved Orrery for mean Motions," of which the former has been lost, and the latter is extant. The orrery for equated motions was initially proposed to be constructed at the Royal Institution, but the plan failed, and the machine was eventually built by the clockmaker John Fidler around 1806. This huge machine, with a complicated gearing mechanism, was a combined tellurian and lunarium. The terrestrial globe was 9 inches in diameter, with a lunar orbit 8 inches in radius, located 14 inches from the central Sun, and all the above parts were mounted on a mahogany table that was 30 inches in diameter. It could present effects of lunar precession: both the Earth and Moon moved in their correct elliptical orbits, and all three of the Moon's periods—the synodic, draconitic, and anomalistic months—were accurate.[49] This orrery was also the very model depicted in the group portrait of Pearson's family on display at the Royal Astronomical Society. Unfortunately, the masterpiece has not survived. Another example of Pearson's improving planetary machines was the mean-motion orrery, a smaller model compared to the orrery for equated motions. The mean-motion orrery, constructed by John Fidler circa 1813, was able to represent the actual mean motions of planets. It is the only surviving planetary machine devised by Pearson and is on display at the Science Museum in London (museum inventory no. 1950–55).[50]

Compared to Pearson's established status within London scientific circles, John Fulton was a relatively obscure country craftsman. He came from a much humbler background as a cobbler by trade in Fenwick, Scotland.[51] Fulton's amateur fascination with astronomy led him to construct orreries between 1823 and 1833. He studied the layout of orreries in the publications of Ferguson and Pearson. Fulton's third and final model was an intricate large machine: an enlarged version of Pearson's mean-motion orrery, 1.37 meters high, with its longest arm carrying Uranus measuring 1.32 meters.[52] Fulton completed this large orrery in 1833 and arranged a tour to exhibit it around Great Britain in Glasgow, Edinburgh, Carlisle, Liverpool, Manchester, and eventually London. The orrery continued to be exhibited in London and Glasgow after Fulton's death. Impressed by the orrery's educational value, it was bought by the City of Glasgow and is currently on display at the Kelvingrove Art Gallery and Museum.

Astronomers Are Not Amused

Pearson's and Fulton's designs represented a particular direction for the improvement of orreries toward achieving a more scientifically accurate representation of planetary motions through more precise calculations and more complex gearing. Not all orrery reformers took this direction, nevertheless. Some, especially those who lectured onstage, focused on the machine's visual performance rather than rigorous scientific accuracy. The transparent orrery invented by the Walkers and similar ones later used by George Bartley and C. H. Adams were probably the most intriguing yet mysterious orreries of this kind.

In chapter 1, I discussed the rise of Adam Walker and his heir William Walker in the late eighteenth century. Adam Walker started devising the transparent orrery while Benjamin Martin was proposing the design of a modular configuration of the orrery with separate attachments in the early 1770s. The resulting device contrived by Walker was the eidouranion, an elaborate machine officially presented circa 1782, which stood vertically in front of the spectators, with a diameter of 20 feet, hence large enough to be clearly seen by everyone in a theater. Walker's gigantic planetary machine and many of its ilk once enjoyed considerable popularity in nineteenth-century theaters, but today very little is known about their mechanisms. Their sizes instantly dwarf Fulton's Orrery (about 6 feet across) and Facey's vertical model at the Crystal Palace (9 feet across). The eidouranion underwent several expansions by William and D. F. Walker, eventually reaching 27 feet in diameter by 1824. The orrery used in Bartley's lecture was described as "a circle of one hundred and thirty feet," which was therefore about 41 feet in diameter and perfectly matched the slogan "Magnificent orrery of unparalleled extent" on the playbill.[53]

Despite their unusually large dimensions, none of these machines are known to be preserved. We can only find clues in scarce sources containing some fragmentary descriptions and very few illustrations, which offer little information about their mechanical details. The likes of the transparent orrery were once favorable in British theaters and lecture halls during its golden age in the first half of the nineteenth century, yet it declined and vanished quickly afterward. In the history of geared astronomical mechanisms, they became dinosaurs that left no fossils to be excavated.

We do not know why so little information on the mechanical details of the transparent orreries survives. It might be related to commercial confidentiality: lecturers (who often claimed the originality of their apparatus) did not wish to disclose details of the mechanism to their competitors. Therefore, the pamphlets or tracts in which lecturers advertised their apparatus and courses often became the primary source of the mechanical details. Publishing syllabi or treatises to promote the lectures was a common practice of eighteenth-century itinerant lecturers, and some nineteenth-century lecturers continued the practice. Such publications were often printed for and sold by the lecturers themselves.

As the pioneers who first introduced the transparent orrery, for example, the Walkers' tracts on the eidouranion lecture were the most important among these sources. The Walkers published *An Account of the Eidouranion*, also titled *An Epitome of Astronomy* in later editions, in 1782 and its thirty-first edition came out in 1824. This series of tracts shows the eidouranion as a family franchise run by the father and sons. Before the twenty-sixth edition in 1817, the tracts were published under the name of William Walker until his death. Deane Franklin Walker, William's youngest brother as well as the heir to the lecturing business, became the author of later editions. In any case, the brothers' father was much respected—all editions credited Adam Walker on the title page as the original inventor of the eidouranion. *An Epitome of Astronomy* was a booklet consisting of around forty pages in duodecimo or octavo and usually contained a notice of the curriculum in the back matter. It was a synopsis of the eidouranion lecture including every scene. The Walkers were nonetheless not explicit about technical details of the eidouranion in this official guide to the lecture. They merely indicated the originality and peculiarity of the eidouranion, which was designed "to give a more natural and comprehensive view of the celestial phenomena than any mode hitherto attempted." As D. F. Walker described the machine in the last edition of *An Epitome of Astronomy*: "This elaborate Machine is twenty feet high, and twenty-seven feet diameter: it stands vertically before the spectators; and its globes are so large, that they are distinctly seen in the most distant parts of a Theatre. Every Planet and Satellite seem suspended in space, without any support; performing their annual and diurnal revolutions without any apparent cause."[54] The above descriptions were almost identical to the one penned by William Walker in the earlier editions around three decades earlier except for the size of the eidouranion, which had increased from 20 feet to 27 feet over the years. D. F. Walker believed that the machine "certainly approaches nearer to the magnificent simplicity of nature, and to its just proportions of magnitude and motion, than any Orrery yet made." He also highlighted the eidouranion's instructive value and aesthetic appeal, proclaiming the machine as a brilliant and beautiful spectacle that conveys to the mind the most sublime instruction as well as rendering astronomical truths intelligibly. Even those who "have not so much as thought upon the subject" may "acquire clear ideas of the laws, motions, appearances, eclipses, influences ... of the planetary system."[55] Walker also stressed the artistic quality of the eidouranion, as he reminded readers that the design of one of the scenes was created by a Royal Academician and the painting was executed by the finest artist in London.[56]

The Walkers' account of the eidouranion in *An Epitome of Astronomy* was skeletal partly due to the fear that imitators would steal their design, a customary practice among astronomy lecturers during the time. George Bartley, C. H. Adams, and many other showmen also did not disclose details of their staged planetary machines in their manuscripts or advertisements. In the *Ouranologia*, in which the producer Samuel James Arnold explained Bartley's lectures to the Lord Chamberlain, Arnold described the synopsis and splendid effects of each scene, but not *how* the machinery behind the scenes would work. Another example was Ebenezer Henderson, who briefly described his invention "Astronomion" in the back matter of his treatise: "THE ASTRONOMION is an astronomical machine on an extensive scale, invented by the Author for his Public Astronomical Lectures; it measures about thirty feet in length, by twelve feet in height, the two sides of which are occupied by diagrams; the central part is for the display of STATIONARY and REVOLVING TRANSPARENCIES, DISSOLVING VIEWS, &c., each of which is about TEN FEET in diameter, and set in motion by clockwork. To this machine, and the Astronomical Lectures, the attention of Literary, Scientific, and Mechanics' Institutions is particularly requested."[57] Like the eidouranion, Henderson's Astronomion has not survived and leaves no trace except for this short descriptive account. The Astronomion seems to be an improvement of the eidouranion, since Henderson showed a list of the "revolving transparencies" for displaying multiple astronomical subjects, including a transparent planetarium (Copernican system of planets), a transparent orrery (focusing on the Earth, Sun, and Moon), a transparent cometarium (exhibiting the motion of comets around the Sun), and a transparent "retrogradarion" (showing the retrograde motions of Venus and Mercury). Furthermore, the Astronomion also provided another fifty "stationary transparencies" of astronomical diagrams and celestial scenery of planets.[58]

It is unlikely that the likes of the eidouranion and the Astronomion were merely devised to enlarge an ordinary orrery and to make it vertical. What exactly are the eidouranion and other planetary machines for the purposes of stage shows? Scholars still disagree on this issue. Some historians of the magic lantern, such as Mark Butterworth and Wendy Bird, presume that the eidouranion was actually a magic lantern show.[59] The historian of astronomical clocks and planetariums Henry C. King, in contrast, speculates that the construction of the eidouranion appears to be a modified and enlarged version of Huygens's planetarium, presumably a geared machine concerned with scenic effects rather than accuracy of wheelwork.[60] Alternatively, the eidouranion might be a hybrid between optical and mechanical devices. Jan Golinski suggests that it was probably a lantern slide with moving parts, similar to the mechanism of "Keevil's mechanical lantern slide" preserved at the Science Museum in London (museum inventory no. 1902-104).[61] This surviving item is said to have been used by an itinerant lecturer, Mr. G. M. Keevil, around 1838, but there is no firm connection between the eidouranion and this device.

Many clues hint that the eidouranion and other transparent orreries were actually hybrids combining clockwork and transparencies. Two feature articles in the *Magazine of Science* offered the most straightforward and detailed contemporary accounts of the mechanism of these "astronomical illustrations" commonly seen in astronomical lectures. The *Magazine of Science*, edited by the science writer George William Francis, aimed at presenting a plain explanation of the principles of natural phenomena, curious inventions, and everyday arts and crafts.[62] The first feature article on astronomical illustrations shows a diagram

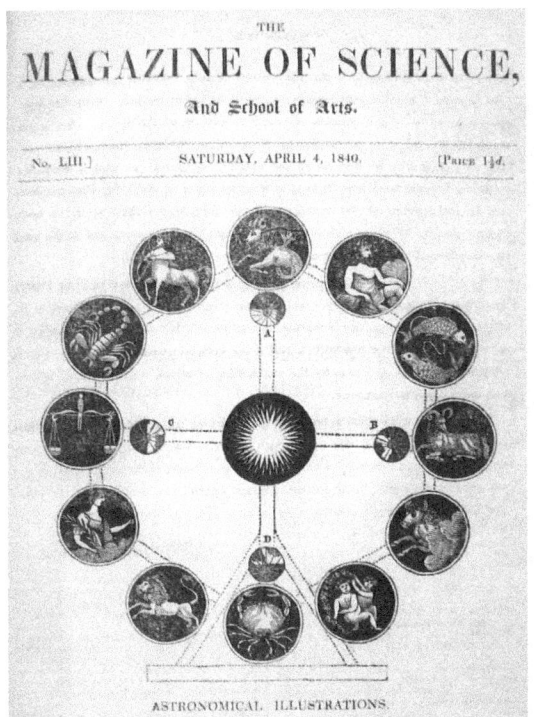

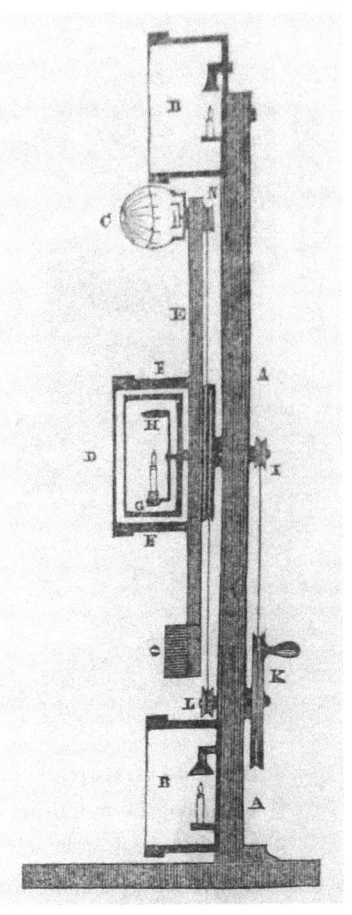

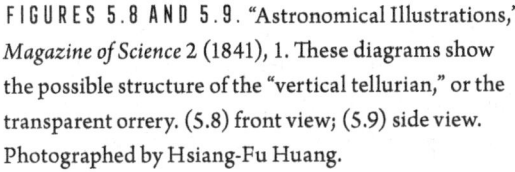

FIGURES 5.8 AND 5.9. "Astronomical Illustrations," *Magazine of Science* 2 (1841), 1. These diagrams show the possible structure of the "vertical tellurian," or the transparent orrery. (5.8) front view; (5.9) side view. Photographed by Hsiang-Fu Huang.

of a "box for exhibiting astronomical transparencies," which explains the lantern slides and box devices used by Mr. Wallis, who was "certainly the best of our astronomical lecturers," presumably John Wallis.[63] The second article introduces diagrams of the "vertical tellurian," a Ferris-wheel-style machine demonstrating the motions of the Earth and Sun surrounded by a circle of transparent signs of the zodiac (figs. 5.8 and 5.9). The journalist was pleased to present this exclusive account of the enormous apparatus, as he believed that "no work whatever contains a plain description of any of these machines." The diagram of the vertical tellurian was represented by a simple arrangement of wheels and cords, though the journalist indicates the machine could be driven by cogged wheels, with the same effects, and "Mr. Wallis's ma-

FIGURE 5.10. Mechanical lantern slide of the solar system, made in 1838 and probably used by Mr. G. M. Keevil. Science Museum, inventory no. 1902-104. © The Board of Trustees of the Science Museum, London.

chinery is a mixture of both, chiefly the latter." The article also specifies the scale of the machine: the zodiacal signs, for instance, should be "at least 3 feet diameter," while the Sun is 2 feet and the Earth is 18 inches. The entire machine could be "as large as the room" and "for theatrical exhibition 30 or 40 feet is not unusual."[64] Although the article did not mention the Walkers and the eidouranion, the form of the machinery was very similar to the illustration of the exhibition of the eidouranion in the English Opera House (see ch. 2, fig. 2.4).

The news report of the vandalism of C. H. Adams's apparatus, mentioned in chapter 3, was another example revealing the hybrid characteristic of the transparent orrery. In the incident, the intruders raided the theater to "abstract the glass of the lantern, break in the face of the orrery's sun, knot up all the cords." This description showed that Adams's orrery could be broken physically and that it contained cords, while the lantern was also included in the machinery.[65] The introduction of Walker's eidouranion and Bartley's lecture in *London Lions* also begin with the remark, "Of those optical exhibitions of the higher class . . . which have been perfected by the aid of painting and mechanics . . . "[66] All the above accounts suggest that the eidouranion and the like were a stage set combining optical and mechanical devices, consisting of cords, cogs, wheels, transparent globes or glasses, lantern transparencies, and light sources. The gigantic visual effects displayed by the transparent orreries were unlikely to have been achieved through the optical projection directly produced by magic lanterns.

Keevil's mechanical lantern slide at the Science Museum (fig. 5.10), as Golinski suggests, corresponded to a hybrid mechanism driven by cogged wheels, as described in the *Magazine of Science* article. However, as the object measures a mere 26 centimeters long and 18 centimeters wide, its scale hardly matched that of the eidouranion, which could fill the entire room. Even if the mechanism could be enlarged, the machinery would be difficult to transport, assemble, and operate smoothly. Keevil's mechanical lantern slide and its ilk were therefore only suitable for a lecture with a small audience, for an itinerant lecturer who had to travel to many locations, or for use in a domestic environment. They were more likely a miniature alternative of the eidouranion, rather than an identical model with the same mechanism.

Another exquisite example of a scaled-down model of the eidouranion was Thomas Elton's "Miniature Transparent Orrery," which was indeed dedicated to D. F. Walker and his family by its maker (fig. 5.11).[67] The device was a simplified tabletop version of the famous eidouranion. Its structure was quite simple: it was basically a case with a glass window in the front. There were two rollers inside the box, around which was wound a canvas scroll with hand-painted scenes. When rolling the reels to turn the rollers, the reproduction of the eidouranion scenes could be displayed one by one. The back of the box could be opened to allow a light source to illuminate the translucent scenes. Of course, the canvas scenes were static, without the revolution of the heavenly bodies or other dynamic effects of the real eidouranion. The principle of the device was similar to today's scrolling light boxes for advertisements. A label on the case read: "To Deane F. Walker Esq., himself and family having by their lectures diffused a general taste for astronomy." The date on the label was February 1817; Elton also created advertisements for the device in several newspapers before and around the beginning of the year.[68] We do not know whether D. F. Walker approved of this "dedication" and the merchandise of a toy model of his family's legendary property. At first glance, Elton's merchandise was likely his own exploitation of the

FIGURE 5.11. Miniature transparent orrery made by Thomas Elton in 1817. History of Science Museum, Oxford, inventory no. 52003. © History of Science Museum, University of Oxford.

famous eidouranion for making a profit. However, since D. F. Walker also took over his late brother's lecturing business in 1817 and began giving the same lecture during Lent of the same year, this act was also possibly a joint effort by Walker and Elton to capitalize on the Walker family's long reputation, as a kind of publicity stunt. Regardless of the situation, as Golinski rightly indicates, Elton's Miniature Transparent Orrery "testifies to the impact of the Walker family's presentations on audiences," for some people were willing to replicate a miniature version of the lecture in their homes.[69]

Despite the suggestion of popularity in the above examples, stage astronomy was not always well received by everyone. The commercial and aesthetic achievements of the lectures by the Walkers, Bartley, and C. H. Adams in blending education and amusement together by applying the transparent orrery were the best in the trade. Some other lecturers' showmanship and the qualities of the apparatuses they used were nevertheless poor, as shown by audience experience in the reminiscences of some celebrities such as Charles Dickens and John Hollingshead (to be further discussed in chapter 6). Some experts of science also criticized the transparent orrery for failing to provide accurate scientific information.

William Pearson was one of the most vocal critics of the transparent orrery. In his contributions to several encyclopedias, Pearson explicitly elaborated on and was very critical of the designs of different models of orreries, whether old ones or the latest constructions. However, he largely neglected to mention the transparent orrery. In Abraham Rees's *Cyclopaedia* (serially published between 1802 and 1819), Pearson simply explained why he excluded several famous cases including the eidouranion, writing, "and with respect to the pendulous orrery of Rittenhouse in Philadelphia, and to that lately exhibited at the Pantheon, London, as well as to Mr. Walker's eidouranion, and Mr. Lloyd's dioastrodoxon, we consider these not as objects of close examination, but as conveying only general information by a scenic effect, not depending on the accuracy of the wheelwork, and therefore not claiming our minute attention."[70] Pearson also justified his decision in the *Edinburgh Encyclopaedia* (1808–30), stating that his aim was to present readers with an account of the machinery that was "equally calculated to amuse the learned and to instruct the learner." Although he did not wish to bring "those scenic representations of the heavenly bodies, which are produced by moving transparencies" into disrepute, he deemed them to be "for the amusement rather than the instruction of a wondering audience."[71]

Considering that the editor of the *Edinburgh Encyclopaedia* was David Brewster, a prominent figure in the elite circles of metropolitan science in both London and Edinburgh, Pearson's negative opinion on the transparent orrery was likely not unusual among men of science. John Herschel also referred to orreries as "those very childish toys" in *A Treatise on Astronomy* (1833), though his criticism was perhaps about orreries in general rather than the large transparent ones. Herschel indicated that to get correct notions of the scale of the solar system through orreries is "out of the question."[72] Both Herschel and Brewster later served as jurists for the Great Exhibition, so it was no surprise that the orreries on exhibit in the Crystal Palace received such harsh criticism. The transparent orrery could not satisfy the learned minds of those scientific elites. The iconic apparatus of stage astronomy was precisely the model Pearson tried to expel from the territory of planetary machines.

Magic Lantern, Movable Slides, and Scenic Transparencies

Geared machines were not the only way to present celestial wonders; the craft of using light was another tool of the trade to delude and please spectators' eyes. As Michael Faraday's remark on William Walker's lecture that astronomy could be illustrated in the most striking way by artificial light, renditions of optical effects became an important part of popular astronomy lecturing as Walker's theatrical performance perfectly demonstrated.[73] Various pictorial entertainments, from small-size shows such as peepshows, to grand spectacles like panoramas depicting the bird's-eye view of the city, used artificial or natural lighting to achieve ideal visual effects. Such a vibrant amusement marketplace in the contemporary metropolis provided incentives as well as competitors for astronomy lecturers to make their exhibitions visually impressive.

Two representative examples of optical displays in nineteenth-century amusements were the Diorama and the magic lantern. The Diorama, first developed by French artist Louis Daguerre in Paris in 1821 and introduced in London in 1823, was a meticulously constructed theater of huge transparent paintings. By elaborate control of skylights and windows behind the paintings, the theater could produce naturalistic lighting effects, thus deluding spectators into thinking that they were seeing life-size three-dimensional scenery changing with time. The Diorama at Regent's Park, the first of the kind in Britain, achieved immediate success.[74] Visitors to London such as Krystyn Lach-Szyrma described the astonishingly immersive experience of watching the show.[75] Many imitators and other less relevant shows with the same name followed the sensation. Richard Altick points out the fast disintegration of the initially very specific meaning of "diorama" in the exhibition businesses: various pictorial exhibitors chose the word in their usage, so the "dioramas" in the mid-nineteenth century could be different scenic shows and the term lost its specificity.[76]

The magic lantern, the forerunner of a modern-day slide projector, was another important example of optical displays. The magic lantern was in fact an important branch of astronomical visual aids other than the mechanical orrery, and it had a far longer history than modern orreries. Optical knowledge and related techniques such as the camera obscura had already developed before the sixteenth century. Historians consider the Dutch polymath Christiaan Huygens to be one of the inventors of the early magic lantern. By using light sources and lenses to project the pictures on transparent plates, the magic lantern could create vivid and (especially to early spectators) astonishing visual effects. It was a medium widely developed for entertainment and educational purposes in the eighteenth and nineteenth centuries.[77] For a long time people were fond of the frightening, mysterious, and entertaining optical illusions produced by the magic lantern. It also had many variant applications in shows, such as the phantasmagoria, a set of dramatic projections of ghosts and grotesque figures. Like many magic lantern shows, the phantasmagoria depended on auxiliary techniques for making sensational effects before the spectators such as rotating chromatropes and moving slides, combined with the lanternist's showmanship. The popularity of the phantasmagoria even produced guides such as *The Magic Lantern: How to Buy and How to Use It* (1866), instructing readers on how to use magic lanterns to make spectacular dissolving views and to raise a "ghost."[78]

Displays of scenic transparencies were prevalent in the repertoire of nineteenth-century popular astronomy lectures. These scenic displays were often made by the image projections of lantern slides. It is uncertain who first applied lantern projection in astronomical lectures or when; the application had likely been developed before the beginning of the nineteenth century. Various sources indicate different origins and there were certainly separate developments in different countries. For instance, some of the earliest known commercial astronomical lantern slides in continental Europe date from around the end of the eighteenth century.[79] In Britain, according to *London Lions*, the optician and artist Charles Blunt was arguably the first person to introduce the lantern technique into popular astronomy lectures. *London Lions* claims that the figures of constellations and the telescopic views of the planets were first painted on glass for the magic lantern exhibitions around 1800, and that Mr. Blunt also originated and executed the idea.[80] Another contender for the title was the instrument-making firm W. & S. Jones, whose 1797 catalog advertised "A new set of moveable painted sliders, shewing the fundamental principles of astronomy, with the real and apparent motions and positions of the planets."[81] This new set of astronomical slides, accompanied by an improved lantern, cost £12/12/0, slightly more expensive than a complete set of the Jones-type portable orrery they sold at the same time. This was the earliest record that lantern slides related to the topics of astronomy had been mentioned in W. & S. Jones catalogs. Nevertheless, William and Samuel Jones channeled most of their energies into the business of making and trading scientific instruments, rather than presenting public lectures on astronomy. In any case, the above accounts may indicate that astronomical lantern slides were introduced in the market before D. F. Walker revived the eidouranion in the late 1810s.

However, the widespread use of lantern slides in public lectures was not common until the invention of limelight in 1825. Before the invention of limelight, the magic lantern relied on candles or oil lamps as light sources, power sources that were far from ideal. By heating a piece of lime or calcium oxide with oxygen-hydrogen gases a flame was produced; limelight provided a powerful, stable source of illumination, which upgraded the use of the magic lantern as well as other optical instruments for public demonstrations.[82] By the mid-nineteenth century, the use of lantern slides in astronomy lectures had become commonplace. Ebenezer Henderson's *A Treatise on Astronomy* (1843) remarks that popular astronomy lectures of recent years "have been rendered very attractive by the introduction of improved transparencies, stationary and revolving, produced by the phantasmagoria."[83] What Henderson meant by "phantasmagoria" here was not the specific shows of ghost and grotesque figures, but a rather general designation for those movable or static projections made by the magic lantern. Advertisements for lectures by C. H. Adams or George Bachhoffner also proclaimed that the lectures would be illustrated by "transparent moving scenery" or "dissolving diagrams," which adopted intricate sets of the magic lantern and dissolving-views apparatus.[84] Newspaper reports also frequently mentioned the use of phantasmagoria by astronomical lecturers; some lecture publications were even accompanied by a box of commercial hand-painted lantern slides aimed at domestic use or the elementary teaching market.[85] This trend was not limited to itinerant stage lecturers or entertainers, but also spread to some much more serious collegiate lecturers. Robert Ball, the foremost astronomical professor-popularizer in Britain after the 1870s, was famed for the extensive use of lantern slides in his lectures.[86]

Astronomy was a popular subject for educational lantern slides. The subjects depicted on the lantern slides reflected the topics that concerned lecturers the most and the effects they expected to present to the audience. The Sun, Moon, tides, eclipses, comets, planets, solar system, and historical cosmological models such as Ptolemaic and Tychonic systems, were common subjects on the lantern slides. Martin Bush categorizes the imagery of nineteenth-century astronomical lantern slides into the "geometrical mode" and the "evocative mode." The former includes observations of celestial bodies, geometrical expressions of cosmological systems and diagrams to explain physical laws. The latter focuses on the spectacles of astronomy through depictions of place-centered scenery, whether an actual or imaginary landscape, to evoke feelings of a meaningful relationship between nature and humans.[87] These two modes of visualization co-constituted visual representation in nineteenth-century popular astronomy in the marketplace.

Despite the popularity enjoyed by the Walkers, Bartley, Adams, and Bachhoffner, the transparent scenes that these lecturers presented left very few extant images and only textual descriptions of them have survived. What did transparent scenes in these early nineteenth-century popular astronomy lectures really depict? Some commercial products related to astronomy lectures provide visual evidence for us to reconstruct the grandeur of the magnificent scenery of stage astronomy. Elton's Miniature Transparent Orrery, introduced in the previous section, was a fine example of the reproduction of the eidouranion scenes. Other kinds of common merchandise included mechanical lantern slides, or "moveable astronomical sliders," as contemporary instrument makers called them. These devices were similar to Keevil's mechanical lantern slide but were smaller and simpler. Each slide, such as the one preserved at the National Maritime Museum

FIGURE 5.12. One of the diagrams from a set of astronomical lantern slides made by Newton & Company, circa 1850. This diagram demonstrates the cause of the tides on the Earth. National Maritime Museum, inventory no. AST1098. © National Maritime Museum, Greenwich, London.

in London (museum inventory no. AST1098), was around 24 centimeters long and 12 centimeters wide, could be held in one hand and manipulated (fig. 5.12).[88] The translucent images on the gearing could be illuminated by outer light sources and stood out in darkness. As the operator turned the handle, the tiny images of the celestial bodies moved, demonstrating astronomical phenomena as the onstage large transparent orrery did. This example in the National Maritime Museum's collections was a set of nine slides, contained in a wooden box, produced by the firm Newton & Company around the mid-nineteenth century. Similar sets of astronomical slides were made and sold by numerous companies at the time, and they were often accompanied by lecture outlines. A single lantern slide usually cost 5–10 shillings, and a set of slides ranged from £2 to £10, depending on the number of slides in a set and the quality of the workmanship.[89] These miniature devices were sometimes a suitable replacement for the large transparent orrery: they could be offered as souvenirs after the lecture, playable gadgets to entertain guests in parlors, and lightweight teaching aids at home or in school classrooms.

Prints of astronomical diagrams and illustrations, whether bound or as a single sheet, were common as merchandise in the popular astronomy market. One extraordinary example, the illustrated volume *The Beauty of the Heavens*, provides a brilliant visual account for present-day readers allowing them to imagine the magnificence of scenic transparencies in astronomical lectures. First published in 1840, this quarto volume contained 104 scenes of full-page color illustrations. *The Beauty of the Heavens* was evidently popular because it reached a fourth edition in 1849. These sales were impressive, particularly considering that the book's price of 2 guineas was not cheap.[90] According to the title page, the author of *The Beauty of the Heavens* was Charles F. Blunt, a lecturer on astronomy and natural philosophy about whom very little biographical information is known.[91] Perhaps this Charles F. Blunt was the very Mr. Charles Blunt who was credited with the first use of lantern slides in astronomical lectures by *London Lions*. This theory is strengthened, despite no direct evidence, by another earlier illustrated astronomical work *Uranographia* (1836), penned under the name of Blanche Elizabeth Blunt. Blanche Elizabeth was Charles F. Blunt's daughter, who acknowledged the good fortune of an early rudimentary education of astronomy she received from "a talented parent" in the preface of the *Uranographia*.[92] *The Beauty of the Heavens* was a more comprehensive sequel to the *Uranographia*: the first section of the former was a barely revised version of the latter; therefore, the *Uranographia* can be regarded as the pilot project for *The Beauty of the Heavens*. It is unclear why Charles F. Blunt would name his daughter as the author of a pilot publication; it was perhaps related to his bankruptcy at the inconvenient time of publication of the *Uranographia*.

The structure of *The Beauty of the Heavens* was close to that of a contemporary popular astronomy lecture. C. F. Blunt asserted that "the Illustrations form the miniature scenery of a public exhibition, such as is occasionally witnessed in lecture-rooms; the text presenting the substance, the order, and the actual delivery of what becomes, in the present instance, a FAMILY ASTRONOMICAL LECTURE." He also believed that a family "need not henceforth quit their own parlour, or drawing-room fireside, to enjoy the sublime 'beauty of the heavens'" with the assistance of his illustrated volume.[93] Therefore, even if this book was not a direct account of a particular lecture, we can perceive it as similar in quality.

The illustrations in *The Beauty of the Heavens* are aesthetic as well as functional. Scientific truth is important, yet adequate enhancement is equally necessary. Blunt stressed that these illustrations had been carefully executed from original drawings, paintings, and observatory studies. Though these figures were "added, occasionally, by appropriate pictorial embellishment," but they were still done "with strict adherence to fidelity of detail." He ensured readers that "great pains have been taken to insure accuracy alike in its pictorial and scientific departments."[94] Several scenes could be seen as counterparts to the familiar subjects appearing in Walker's, Bartley's, and Adams's lectures. For example, Scene no. 2 is a classic diagram showing the appearances of masts on an inbound or outbound ship to prove the Earth's round shape.[95] This scene was included in the lectures of both Walker and Bartley; the only difference between the images in Blunt's book and in an actual lecture was that the former was a static diagram, while the latter was a demonstration with a model ship moving around a globe. Other explanatory diagrams such as "The Phenomena of the Seasons" and "The Phenomena of the Tides" were common topics in contemporary astronomical lecturers' syllabi.[96] These diagrams correspond to the category of the geometrical mode in Bush's analysis.

On the other hand, many scenes in *The Beauty of the Heavens* belong to the category of the evocative mode. They were designed to evoke feelings of the sublime through drawings of beautiful landscapes or cosmic spectacles. The telescopic view of individual planets, the classical illustrations of the celestial hemispheres and constellations, and landscape drawings containing astronomical phenomena were in this category. For example, Scene no. 58, "Eclipse of the Moon," shows the eclipse in the background of a town beside a smoking volcano resembling the city of Naples with Mount Vesuvius (fig. 5.13).[97] In the foreground are three small figures, presumably amazed, though we do not see their facial expressions, witnessing and pointing their fingers at the eclipse. The entire scene depicts the heavenly spectacle in a beautiful, serene, and romantic sense. Overall, the illustrations in *The Beauty of the Heavens* were elegantly rendered by the artist. If the plates in an illustrated book could evoke

FIGURE 5.13. "Eclipse of the Moon," Scene no. 58 from *The Beauty of the Heavens*, by Charles F. Blunt (author), London, 1842, 79. Image digitized by ETH-Bibliothek Zurich.

wonder in readers' emotions, it is not surprising that the audience would feel awe when seeing a more magnificent representation on the stage.

Orreries and numerous variants of devices evolved from them had been important apparatuses for popular astronomy lectures since the eighteenth century. These apparatuses not only functioned as visual aids to plainly demonstrate the movements of celestial bodies and the causes of astronomical phenomena. At first, they were also deemed instruments of enlightenment science par excellence and later developed into devices for showing the wonders and grandeur of the universe, hence becoming symbols of the sublime science. Even though the improvement of orreries had stalled by the mid-nineteenth century and their use was no longer considered serious and advanced by men of science, the orrery remained an icon of astronomical studies for the general public (perhaps second only to the status of telescopes). A full-

FIGURE 5.14. Illustration of the orrery from *Smith's Illustrated Astronomy*, Asa Smith, New York, 1855, 6. Image digitized by the Library of Congress, Washington, DC.

page illustration at the very beginning of a textbook, *Smith's Illustrated Astronomy* (1855), best exemplified this iconic cultural status of the orrery in astronomical education (fig. 5.14).[98] In the lower center of this illustration, a generic orrery, though unrealistically large, stands on the ground and corresponds to the diagram of the solar system in the sky above. Two teachers are instructing a group of students in the structure of the solar system, while a pupil is operating the larger-than-life orrery. Orreries had become the instruments that people first thought of when it came to teaching the rudiments of astronomy.

Many nineteenth-century scientific elites were very critical of, or even expressed contempt for, those "childish toys" encompassing various types of orreries. As I have shown in this chapter, however, some were still unable to deny the educational value of these visual aids. Lantern slides were especially promising among them as optical projection devices that had developed quickly from merely frivolous entertainments to powerful tools in education. David Brewster, for instance, held a more favorable opinion of the magic lantern, despite being a member of the juries at the Crystal Palace who had severely criticized the orreries on exhibit. Brewster endorsed the value and importance of the magic lantern in the handbook *The Magic Lantern*, asserting that the apparatus for "amusing children and astonishing the ignorant" could also contribute to a visual education.[99] Robert Ball, a scientific celebrity, the Victorian counterpart to Carl Sagan, extensively used lantern slides in his astronomical lectures. Ball's meticulous attitude toward lantern slides shows in his having collaborated with several lanternists to make particular slides and even preparing instructive notes to specify his requirements.[100]

Despite their decline in use, some mechanical orreries such as the Jones-type model still outlived the fashion of Lenten astronomy lectures and were manufactured until the early twentieth century.[101] In the frontispiece of Ball's bestseller *Star-Land* (1889), he was shown giving a juvenile lecture at the Royal Institution. A tabletop orrery bearing a resemblance to a Jones-type model was put on the table in front of him, and a large magic lantern was in the background (fig. 5.15). This image is clearly reminiscent of continuity with the tradition of scientific lectures in popular astronomy lecturing. From Ball to Sagan, visual aids like the orrery and lantern slides have always been necessities in demonstrations of the heavens.

FIGURE 5.15. "A Juvenile Lecture at the Royal Institution," frontispiece to the *Star-Land*, Robert Ball, London, 1889. © The History Emporium / Alamy Stock Photo.

CHAPTER 6

AUDIENCES

Readers did not play merely passive roles in the nineteenth-century popular science publishing trade. Feedback from readers could influence the decisions of the editorial staff. The "invisible hand" of readers, for example, was noticeable in the editing process of the Religious Tract Society's popular science publications. The editorial department of the RTS was assisted by a group of in-house readers and external expert readers. Manuscripts were first sent to an internal reader and subsequently to an external one. The two readers had different tasks: the latter, an expert on related subjects, could evaluate scientific accuracy, whereas the former was to decide whether the manuscript was written in a suitable style and with an adequate Christian tone. The manuscript would be rejected completely if both readers reported unfavorably on it and the editor saw no reason to disagree.[1] This procedure, like today's peer-review process for publication in academic journals, not only controlled the quality of the publications but also ensured that they would engage the society's intended audience, Christian readers. The RTS was clearly aware of its readership and carefully managed it.

Likewise, the audience for popular astronomy lectures did not consume knowledge and amusement passively. They were actively involved in molding the fashion for astronomy, and lecturers were responsive to the market. Although lecturers conveyed messages concerning the wonders and sublimity of the universe, audiences might interpret the performance very differently. The education or rational amusement offered by a lecturer was often missed by a pleasure-seeking audience. Audiences were customers as well as participants in the science marketplace.

In today's elections or commercial competitions, evaluating voters' intentions or customers' attitudes toward a product pose difficult challenges for decision-makers. That is why the results of various public opinion polls are often as dizzying as riding a roller coaster or looking through a kaleidoscope. Researchers in different disciplines have also acknowledged the complexity of audience composition and reader groups in the media as well as the difficulties of analyzing them. Many approaches have been developed to investigate audience composition, attitude, and reception. In literary and theater studies, reception theory derived from the reader-response concept is a widely applied analytical approach to studying audience behavior. Reader-response emphasizes the relationship between readers and texts. It concerns how readers contribute to the meanings of a literary work—in other words, their interpretation of the text beyond the author's original intention.[2] In the field of science communication, too, various strategies have been discussed for evaluating the effectiveness of communication. Early advocates of the public understanding of science movement in Britain in the 1990s often adopted the "deficit model," the assumption that the lay public lacked scientific knowledge, and a science popularizer's job was therefore to enhance the scientific literacy of an ignorant, passive audience. However, this top-down view was later criticized for being too prescriptive and scientist-centered, and for neglecting mutual communication between popularizers and the general public.[3]

In the history of science, many researchers also employ approaches from cultural history and the history of books to investigate the readership of popular science publications. James Secord's study of the infamous Victorian bestseller *Vestiges* is an impressive milestone in the development of this genre. Secord claims that his work has been an experiment in a different kind of history, for he uses sources from everyday practices, including "diary keeping, letter writing, displaying, debating, lecturing, and conversation," to explore the major historical episode of an evolutionary bestseller and its aftermath from the perspective of reading.[4] Another representative example is Jonathan Topham's work on the eight volumes of Bridgewater Treatises. Topham analyzes different classes of readers from fashionable society to radical artisans. He also demonstrates that readerships were largely shaped by a series of elaborate negotiations between authors, publishers, printers, booksellers, and libraries. The case of the Bridgewater Treatises shows that publishing—as well as the reading of books—is "embedded in a complex and varied series of social relations."[5] In addition to studies of print culture, there are studies of audiences for various events including lectures, exhibitions, and museum visiting. For example, the scholarship on lecturing at the Royal Institution provides detailed analyses of audience size and formation during the early and mid-nineteenth century, especially emphasizing the prominence of Michael Faraday's activities.[6] The methods used in these studies form the basis of my research on the astronomical lecture audience in this chapter.

Audience Composition

Identifying the right target audience is critical for creating a bestseller. A successful writer must think about who will buy the book and leaf through its pages. In the preface to *The Earth's Crust* (1864) by David Page, the author describes the intended audience for his popular geology book. Page suggested that the readers of his monograph would include various groups such as "young men striving after self-instruction," "men in business, whose time will merely permit a cursory acquaintance with a subject," people at their leisure "who seek information simply as an accomplishment," and the "gentler sex, unprepared for technicalities."[7] Page declared that his handy and intelligible outline of geological science was made to meet these readers' needs. From Page's claim, we can easily get a rough, though subjective, picture of the constitutions of his book's readership.

It is never an easy task to trace the readership of a book. Studies of public scientific lecturing encounter the same problems in mapping audience composition. In figure 6.1, a renowned oil painting depicting Michael Faraday's Christmas lecture at the Royal Institution, the artist also shows the large audience in the lecture theater. The most conspicuous audience members were Prince Albert and his two sons in the center of the front row; one of the little princes was the future King Edward VII. We can surmise based on the audience's attire that this was a fashionable upper-middle- or upper-class audience. However, apart from Prince Albert, the two princes, and a few of the spectators in the front row and foreground of the picture, most of the audience members have blurry faces or appear only in silhouette, making it al-

FIGURE 6.1. *Michael Faraday Lecturing in the Theatre at the Royal Institution*, by Alexander Blaikley, circa 1856. © The Royal Institution / Science Photo Library.

most impossible to identify individuals. This depiction best represents the situation a researcher encounters when studying an audience or a readership: most of the audience in the background is blurred or even faceless, whereas images of a very few individuals in the foreground are distinct. Historians' understanding of the audience, however, usually comes from this very small portion of attendees in the foreground.

In addition to the scholarship on scientific lecturing, I borrow tools from historians of Victorian scientific publishing or theater studies, which offer many approaches to determining the readership. The most direct sources, while often sparse and difficult to collect, are accounts by the readers themselves: diaries, letters, and memoirs that reveal their personal reading experience. On the other hand, accounts of description or objectives written by the authors or the publishers, such as Page's declaration in his book's preface, are other clues suggesting the potential readership. Another method of surveying readership is to consult the sales figures for books to learn circulation patterns: a book's editions, prices, and print runs are all clues. As James Secord and Aileen Fyfe have shown in their studies of Victorian popular science books, the different styles and qualities of a book's papers, printing, and binding influence the variation of book prices. Editions ranged from hardcover to paperback and even to pirated copies, and they implied diverse economic levels of target customers.[8]

The cases of cheaper editions and potted piracy demonstrate that the availability of a popular science book could reach readers of all social classes. The improvement of literacy and the prevalence of steam-powered printing contributed to a wider range of readers. Although the level of literacy might be disputable, the number of literate people in Britain slowly increased between the late eighteenth and mid-nineteenth centuries. Some statistics suggest that around 60 percent of males and 40 percent of females in England and Wales were literate in the late eighteenth century, and the proportions increased to 69.3 percent and 54.8 percent, respectively, in 1851.[9] Another source shows a generational difference: approximately between 1859 and 1874, the literacy rate for young people in their early twenties was over 65 percent, while that of the middle-aged people in their late forties was less than 50 percent.[10] No matter how statistics are calculated, historians agree that literacy rates increased dramatically starting in the 1840s and throughout the second half of the nineteenth century.[11] The spread of Sunday and day schools, along with the missionary activities of evangelical churches or societies, contributed to this improvement. Steam-powered printing made cheap, mass publications possible. Publishers found a new market for cheap pamphlets, tracts, and periodicals, which reached the middle- and working classes.

Studies of visitors to lectures, exhibitions, and museums employ similar methods, ranging from the direct accounts of audience members relating onsite experiences, to studies of institutional membership, regulations, and attendance records. Admission charges, analogues to book prices, provide researchers with clues about the socioeconomic status of the audience. Historians have used the above methods to survey visitors to the Friday evening discourses at the Royal Institution, the Great Exhibition of 1851, and several Victorian museums of anatomy and natural history.[12] I use many of these same methods to analyze the audiences of popular astronomy lectures. The

AUDIENCES

TABLE 6.1. CATEGORIES OF THE AUDIENCE

Category	Subdivisions	Indications
Age	Children / juveniles Adults (especially those in supervisory roles, such as parents, guardians, and teachers)	ABCD
Gender	Females Males	BD
Social classes / finances	Upper classes Middle classes Working classes	ABCD
Political or religious views	Christians Atheists Political radicals	BC

The main categories of audience covered in the analysis of this chapter. The following indications are employed: (A) Admission charges for the lecture. (B) Personal accounts or reports of the lecture, such as periodical articles and memoirs. (C) Institution's or lecturer's agenda. (D) Visual evidence of the lecture, such as illustrations.

indicators I employ to survey audiences include admission charges, personal accounts, lecturers' or institutional agendas, and visual evidence of lecture scenes (see table 6.1). These methods are nevertheless not perfect and should be treated cautiously. For example, ticket prices could only offer crude, if reasonable, speculation about the social class of the audience. It is a piece of indirect evidence of the possible range of the audience, rather than firm proof of the social status of a specific spectator sitting in the auditorium. Even if a piece of visual evidence of a lecture scene exists, such as the illustration of D. F. Walker's Eidouranion at the English Opera House (see ch. 2, fig. 2.4), it is hard to avoid questioning the depiction's faithfulness. Illustrations might be overly idealized, romanticized, or oversimplified, thus not reflecting the actual scene. Furthermore, even if a picture is realistic, it might depict a special occasion rather than a regular scene, and thus not be representative of an everyday situation. The iconic scene of Michael Faraday's Christmas lecture (fig. 6.1) recorded a special moment when royal family members were present, but did not necessarily represent the institution's everyday events. Visual evidence is therefore powerful but still requires careful scrutiny.

In terms of availability of sources, institutional lectures, especially those given at prestigious scientific institutions, more often preserved everyday records than did private lecturing businesses. The Royal Institution, for example, maintained relatively comprehensive records ranging from lecture attendance to lists of subscribers. This enables researchers to make a prosopographical study, such as one that identifies women subscribers during the institution's early years, as well as precise attendance at a particular lecturer's course.[13] The admission fees for members or subscribers are strong indicators of audience composition. The wide range of subscription fees varying from institution to institution suggests the same broad spectrum of audiences. The objectives claimed by institutions' founders also indicate the specific groups they aimed for, though the original statements might not always be reliable. The goals of the board could change dramatically in confronting financial crisis, forcing these institutions to become entertainment venues rather than places offering more serious education as they had originally intended. Many mechanics' institutes and the Colosseum at Regent's Park encountered similar problems by the middle of nineteenth century.[14]

The audiences for popular astronomy lectures were also heterogeneous, as were many nineteenth-century scientific lectures in Britain, existing across all levels of social strata. The audience that came to Albemarle Street was different from the crowd at the Southampton Buildings of the London Mechanics' Institution. Cheap lectures held at mechanics' institutes were aimed at the working classes. In contrast, the Royal Institution and the like appealed mainly to the middle- and upper classes; its lectures tended to reflect the taste and concerns of the educated elite—if not the landed gentry, at least the professional middle class.[15]

For itinerant lectures held in provincial towns and the countryside, or private lecturing businesses not affiliated with a scientific institution, everyday records might not have been preserved. Nevertheless, researchers can still find clues about the nature of the audience in advertisements, playbills, and syllabi, as well as in data regarding sponsors and venues. For instance, a lecture arranged in a country boarding school, like John Bird's at the public school Charter House under the patronage of the Reverend Dr. Russell as we have seen in chapter 3, the pupils, teachers, and pupils' families were likely to make up the majority of the audience. When the lecturing venue was a theater, the producers usually anticipated a diverse audience, insofar as audiences in theaters during the Regency and Victorian eras often reflected a microcosm of the larger society.[16] As George Cruickshank depicted in a caricature (see ch. 2, fig. 2.2), in the distinctive and hierarchical space of Regency and Victorian theaters, "vulgar" sorts of working men and women shouted, laughed, and even rioted in the top levels of the gallery; the fashionable rich people sat in the private boxes in the middle levels of the auditorium; the lowest pit was often chosen by the middle-class spectators. Despite these separate levels, the different social classes all sat under the same roof.

The prevalence of popular astronomy lectures across all classes of society was also demonstrated in the variety of admission fees. Everyone, whether rich or with just a few pennies in their pockets, could find a lecture that suited their needs and budget. Table 6.2 summarizes admission prices to several astronomical lectures in London. The wide range of these fees suggests that a broad spectrum of society was present in the audience, and shows that managers adopted a strategy of offering affordable prices to spectators of different financial levels. The list indicates that the most expensive ticket for a single lecture was 1 guinea (21 shillings) for a seat in a private box at the Adelphi Theatre to see C. H. Adams. This amount was equal to the average weekly wage of a shopkeeper or a female copy clerk in the middle of the nineteenth century, but was affordable for richer spectators who wanted to enjoy an undisturbed comfortable space.[17] Categories of fees for a lecture in a theater varied depending on the structure and seating plan of the auditorium. In general, the price of a box seat was usually up to 5 shillings, while the cheapest charge for entering the gallery was usually 1 shilling or less. Lectures at scientific institutions would offer more flexible packages such as seasonal or annual subscriptions. Members and their families could also receive admission discounts. Some charitable lectures specifically aimed at the working classes, such as those organized by the radical philanthropist W. D. Saull, as we have seen in chapter 2, even offered for free admission. In contrast, aristocrats or upper-class elites could afford the services of private tutors at their country estates. For example, King William IV and the Duke of Wel-

TABLE 6.2. ADMISSION PRICES FOR POPULAR ASTRONOMY LECTURES

Lecturer	Year	Venue	Prices (Shillings)*
D. F. Walker	1819	English Opera House	5 (B); 3 (P); 2 (G); 1 (UG)
D. F. Walker	1838	Colosseum	3 (B); 2 (Pm); 1 (P)
R. E. Lloyd	—	Haymarket Theatre	5 (B); 3 (P); 2 (G); 1 (UG)
G. Bartley	1822	English Opera House	5 (B); 3 (P); 2 (G)
C. H. Adams	1835	King's Theatre	4 (S); 2 (B); 1 (P)
C. H. Adams	1854	Adelphi Theatre	21 (PB); 10.5 (PB); 3 (S); 2 (B); 1 (P); 0.5 (G)
J. Wallis	1827	London Mechanics' Institution	24 for an annual subscription; 6 for a quarterly subscription
J. Wallis	1846	Royal Institution	Nonsubscribers: 21 (adults) or 10.5 (children) for a course of six lectures Subscribers: 42 for the season Members' children: 21 for all lectures
G. H. Bachhoffner	1845	Royal Polytechnic Institution	1 (adults); 0.5 (schools)
G. H. Bachhoffner	1859	Colosseum	1 (adults); 0.5 (children and schools)

Examples of prices for admittance to an astronomical lecture or a course in London between 1810 and 1860.

Sources: RAS: Add MS 88: 2–8; RI: GB 2: 47; *Athenaeum*, March 5, 1859, 322; *Examiner*, February 15, 1845, 109; *Examiner*, September 27, 1827, 623; *Examiner*, March 11, 1838; and *Morning Chronicle*, March 13, 1835.

*Abbreviations: Box (B); Private Box (PB); Pit (P); Gallery (G); Upper Gallery (UG); Stall (S); Proscenium (Pm).

lington were patrons of John Bird, who occasionally lectured before the royal family at the Pavilion in Brighton.[18]

One audience for popular astronomy lectures were young people: boys and girls, from children to adolescents. Adults have often overshadowed this group of visitors in discussions of audience composition. However, juveniles in fact occupied a significant place in many lecture theaters. Astronomy remained an essential subject in juvenile education by the nineteenth century. This convention had taken root in Enlightenment polite culture, in which the cultivation of astronomy and geography was an important part of the education of gentlemen and ladies.[19] Furthermore, astronomy played a role in natural theology, conveying moral and religious lessons, while at the same time emphasizing the use of human reason. This also commended astronomy to reformists' emphasis on the achievements of human reason. Meanwhile, astronomy was also valuable in navigation and cartography applications. These features made astronomy a worthy subject in the curricula of liberal arts and natural philosophy. Even advocates of mechanics' institutes, which aimed at the working classes, included astronomy in their vision. When the journalist William Cooke Taylor proposed an ideal curriculum for polytechnic schools in 1842, he included astronomy and claimed the subject to be among the most important of the "mixed science[s]."[20]

However, only a small number of young people had a chance to obtain formal primary schooling, much less higher education. Girls had even fewer opportunities than boys.[21] Instead, domestic instruction played a key role in education. Commodities produced for learning astronomy at home ranged from illustrated books to remarkable visual aids and toys, such as the cardboard celestial sphere and the board game.[22] Portable orreries could also be excellent tools for domestic astronomy education in parlors in addition to their public exhibition function. Popular astronomy lectures could serve as a day trip for children at home, or as a pleasant field trip for school classes. Managers and lecturers recognized this aspect, and sought to attract young people to their shows. Thus C. H. Adams and George Bachhoffner offered discounts to encourage parents to bring along their children. Guidebooks such as the *London Lions* also recommended Bartley's and Walker's shows to parents, for "conveying to the minds of their children the lofty and magnificent ideas that Astronomy supplies."[23]

Children, of course, could hardly attend a lecture on their own initiative, and usually did not buy their own books. Parents, guardians, governesses, or teachers were the adults who accompanied and supervised children's learning. A *Punch* cartoon vividly depicted a gentleman dragging a boy through the famous Crystal Palace Dinosaurs at Sydenham Hill (fig. 6.2). Despite the authoritative posture of the gentleman, probably the boy's father or guardian, the boy "Master Tom" seemed reluctant and frightened by the giant prehistoric monsters and "strongly objects to having his mind improved."[24] This caricature demonstrated that children were not always active and enthusiastic audiences for the "improving" of science education. Satirical literature or magazines like *Punch* contained many similar examples lampooning the presumably instructive lectures or exhibitions. These caricatures often poignantly illustrated the disagreements concerning the effectiveness of rational recreation between pleasure-seeking children and their authoritative parents or adult guardians. Another *Punch* cartoon, depicting the misadventures of a lecturer "Uncle Fusby," was representative among these caricatures. The absent-minded Uncle Fusby delivered "delightful" astronomy and chemistry lectures to children, but it was the accidents due to Fusby's carelessness rather than the lectures themselves that caused delight. In the astronomical lecture (fig. 6.3), Fusby was unaware that the diagram and the orrery behind him had been covered in graffiti, and the caption of the cartoon jokingly points out that the lecture was "finally touched up (just before the gas was turned on) by his mischievous nephews."[25]

Because parents and guardians were the ones actually paying the fee and accompanying the children, to aim at the juvenile market also meant to target these adults. "The modern schoolmaster is expected to know a little of every thing, because his pupil is required not to be entirely ignorant of any thing," wrote the English author Charles Lamb, who criticized contemporary schoolmasters for their overzealous instilling of knowledge, even during school intervals. Lamb sarcastically observed that a schoolmaster usually had some "intrusive upper-boy" fastened upon him and must "drag after him to the play, to the Panorama, to Mr. Bartley's Orrery, to the Panopticon, or into the country."[26] This trend toward increasing knowledge and improving the mind was encouraged by guardians and teachers, especially among the upper- and middle classes. Popular science authors and lecturers were aware

FIGURE 6.2. "A Visit to the Antediluvian Reptiles at Sydenham: Master Tom Strongly Objects to Having His Mind Improved," from "Punch's Almanack for 1855," *Punch*, vol. 28 (1855), [8]. © Punch Cartoon Library / TopFoto.

FIGURE 6.3. "Uncle Fusby Undertakes to Delight and Instruct the Young Folk at Christmastime," from "Punch's Almanack for 1866," *Punch*, vol. 50 (1866), [10]. © Punch Cartoon Library / TopFoto.

of this fashion and often advocated their works for family or domestic usage. Charles Blunt, the author of the popular astronomy book, *The Beauty of the Heavens*, claimed that a family could benefit from his illustrated volume. He suggested how this book was to be used: "The Lecture [text] may be read aloud by a parent, teacher, or any member of a party, the Scenes [illustrations] being exhibited, at the same time.... It would be impossible to devise a more rational, or, to a well-regulated mind, a more cheerful mode of passing an evening."[27] A family value blended with communal recreation and domestic education was the core of this appeal.[28] Parents and teachers were therefore an important component of the audience in the popular astronomy market.

The above discussions of juveniles and parents lead us to focus on the presence of women in the audience. Before feminist scholars in the 1980s criticized contemporary scholarship on the history of science and generated broader interest in the neglected women of science, women's participation in science had been often overlooked in research.[29] The achievements of several outstanding women who contributed to studying or popularizing scientific knowledge, such as Caroline Herschel and Mary Somerville, and the difficulties they encountered in their scientific careers, have been the focus of studies ever since.[30] The role of women as patrons, consumers, and audiences of science, however, remains less discussed. Recent studies of BAAS meetings and the early Royal Institution consider women as a collective group in examining women's participation in scientific organizations.[31] In fact, women were prominently represented in most audiences for popular astronomy and other scientific lectures in the early nineteenth century. For example, many contemporary accounts indicated that women made up at least half of the audience at Humphry Davy's chemistry lectures at the Royal Institution. These female attendees could number hundreds at each lecture; many were young ladies or those "of the highest rank and respectability."[32] The regular presence of women participants in the BAAS meetings also reflected a welcome acceptance of women at public scientific events, although this acceptance was based on conventional social and cultural expectations for women. As Rebekah Higgitt and Charles Withers indicate, by participating in public activities of the BAAS such as preparing the town, receiving conference guests, and admiring the knowledge presented by male speakers, women's traditional supporting roles as wives and daughters were extended to support of the BAAS.[33]

The presence of female participants in astronomy, whether as spectators at popular lectures or as active members of local astronomical societies, is also evident. In contemporary illustrations of the scenes of popular astronomy lectures, women are commonly seen in the audience. There was a fashionably dressed female audience for Walker's eidouranion at the English Opera House in 1817 and Mr. Perini's planetarium in London in 1879.[34] The same scenes also depicted children, husbands, or guardian-like male figures. These illustrations suggest the appeal of astronomical shows and their ability to attract a mixed audience of upper- or middle-class families. Allan Chapman's investigation of female members of amateur astronomical societies during the late nineteenth and early twentieth centuries also shows that women were active participants in these communities. Some of these female members were not merely passive members of the audience but enthusiastic observers, writers, and lecturers. Agnes Mary Clerke, for example, was one of the most prominent woman popularizers of astronomy in late nineteenth-century Britain.[35]

Reminiscences of Visiting Shows

Among numerous accounts of theatergoers on the experience of visiting shows, the essay "At the Play," written by the journalist-turned-theater impresario John Hollingshead in the *Cornhill Magazine* in January 1862, is a rare reminiscence of a Lenten astronomy lecture.[36] In this nostalgic and humorous article, an old playbill inspired Hollingshead to recall the plays he had been to as a child. One occasion he remembered vividly was an astronomical lecture, a "treat" offered by a family friend. This family friend was a severe schoolmaster who "objected to theatres upon principle," and yet "saw no harm in going to a playhouse during Passion-week to hear an astronomical lecture, illustrated by an Orrery." Instead of being a play, for the schoolmaster friend such a lecture was "amusement and instruction combined." With the permission of Hollingshead's parents, they went off, full of expectations. The visit probably took place between 1837 and 1840, a few years after Hollingshead's first theatrical experience at the age of nine.

The lecture, unfortunately, did not seem promising even before it started. Hollingshead recollected the bleak space of the theater, with its empty orchestra and silent stage, and the lecturer's apparatuses scattered across a few tables, "like a deserted shop." The auditorium was also deserted: bare benches, empty boxes, and a lack of heating equipment left the few spectators in the pit huddling together for warmth. All these gloomy conditions were unpleasant for the audience, too. Hollingshead described the unhappy spectators: "They were mostly country people, who probably thought they were seeing an ordinary play, or persons who came to perform a solemn duty by learning something about the 'solar system.' If their faces were any guide to their feelings, they looked bewildered and unhappy, except for one individual, who seemed to despise the wonders of the universe."[37] Then the show began. Hollingshead recollected:

> This was the entertainment—amusing and instructing—which my guide had brought me to for a treat. My insolvent theatre, in its most degraded period, was never as dull as this. When the lecturer came on with a jaunty air, and began to patronize, without clearly explaining, the Infinite, I thought I knew his voice and manner, although he was disguised in very clerical evening dress. His style of playing with the Orrery—an apparatus, by the way, which was most creaking and unmanageable—was so like that of a juggler handling the cups and balls, that I . . . soon traced in him [the lecturer] the broken-down manager of my insolvent theatre. I was about to impart my knowledge, with youthful confidence, to my guide, when we were interrupted by a discontented mariner, who had drifted into this unhappy port in search of amusement.
>
> "Hi mate," he said, loudly, to my severe companion, after a number of preparatory grunts, "when's the broadsword combat goin' to begin?"[38]

Things then became truly dramatic. Realizing that there would be no play or "broadsword combat," the sailor shouted at the lecturer to demand his money back. Their quarrel disrupted the show. At last, the grumpy sailor was coaxed out of the theater by a doorkeeper, and the unlucky lecturer, "probably glad of an excuse to hurry through his lecture, professed to be so disturbed by the interruption, that he could hardly tell the sun from the moon." On the way home after the show, Hollingshead's family friend "mourned over the instruction we had been deprived of by a rude boor."[39]

For a piece intended to be humorous, Hollingshead likely chose this

LECTURE IN LENT.

Uncle. "WELL, CHARLEY, SHALL I OPEN ANOTHER BOTTLE OF CLARET, OR WILL YOU GO WITH YOUR COUSINS TO THE LECTURE ON ASTRONOMY?"

Charley. "WHY, UNCLE, I THINK, ON THE WHOLE, I PREFER CORK TO ORRERY." [*Which shows that he was an Epicurean, but had read his* "*Debrett.*"

FIGURE 6.4. "Lecture in Lent," *Punch*, vol. 60 (1871), 105. © Punch Cartoon Library / TopFoto.

extremely awful case and exaggerated the drama. Hollingshead's vivid account still offers us invaluable information about popular astronomy lectures in theaters. The supposedly instructional amusement that the adult guardian regarded as a "treat" was a boring disappointment for the juvenile. The disgruntled sailor was looking for some exciting amusement but came to the astronomical lecture out of a misunderstanding and apparently did not appreciate the education about the heavens. Ironically, his uncivilized manner and ignorance were exactly what the "rational recreation" of Enlightenment polite culture was supposed to improve. We can easily see the dissension within the audience and within generations. There was widespread approval of the proclaimed educational function of astronomical lectures among parents and teachers. As Hollingshead's story shows, even a serious schoolmaster, who objected to theatrical entertainment, could think that visiting a playhouse to attend an astronomical lecture was beneficial to juveniles. This was exactly the educational value that lecturers had advocated and exploited for decades.

A *Punch* cartoon titled "Lecture in Lent" published in 1871, despite being unrelated to Hollingshead's story, also suggests the fashion of bringing youngsters to Lenten astronomy lectures by the mid-nineteenth century (fig. 6.4). In the cartoon, an elder gentleman sits at the dinner table with his nephew Charley, a young lad who appears to be in his early twenties. The uncle asks Charley whether he would like to stay to drink a bottle of claret or "go with your cousins to the lecture on astronomy." Charley replies, with a dry wit: "Why, uncle, I think, on the whole, I prefer Cork to Orrery."[40] The humor of this cartoon is based on wordplay—"Cork" and "Orrery" refer to a bottle of claret and a lecture on astronomy, respectively, but the two words are also both titles in the Peerage of Ireland. Although the cartoon does not directly depict a particular lecture, and its title only sets the "context" of its scene and conversations, it still implies a fashion or convention for adults usually to accompany juveniles or children to attend astronomy lectures in Lent, as Hollingshead's story shows.

The identity of the lecturer and the quality of the lecture mentioned in Hollingshead's article are also noteworthy. The unknown lecturer was recognized by young Hollingshead as the former manager of an insolvent theater. He dressed in clerical clothing on stage. Such dress was a means of self-fashioning as an authoritative figure and reflects the public's perception of what a lecturer on astronomy should look like. As popular astronomy had been commonly associated with religious reflections and the idea of the sublime, it was sensible for a lecturer, especially one unknown or new to the profession, to project an image of authority by dressing like a priest. This is like today's stereotypical impression of a scientist in the media and popular culture, which often depicts a man with messy hair, wearing a white laboratory coat.[41] A former manager of an insolvent theater would try astronomy lecturing as a sideline or an alternative venture. This phenomenon suggests that the popularity of astronomy lectures and the trade was still profitable; on the other hand, it also reveals the loose requirements for becoming an astronomical lecturer. The condition of the apparatus and the theater's facility were as poor as the lecturer's showmanship: the empty orchestra, the creaking orrery, the deserted-shop-like apparatus, and the lack of heating in the auditorium all suggested that the "treat" offered by Hollingshead's family friend was a mediocre cheap show not comparable to the best ones. Anyone could step into the lecturing business whether his previous identity was Cambridge student, journeyman carpenter, or broken-down theater manager, but success in winning the audience was another story.

Hollingshead's recollection was not an isolated case of visiting an astronomy lecture in a theater. Similar childhood experiences appeared in other contemporary accounts. Charles Dickens, for instance, described an orrery visit at a local theater in his childhood. Dickens's visit was an outing for his birthday, probably in Chatham, where he lived until age eleven (1822). However, the birthday treat became "a slow torture," as Dickens put it. The quality of the apparatus and performance was also poor. Dickens described the lecturer as a "low-spirited gentleman;" the instrument, too, was "a venerable and 'a shabby Orrery." During the show the orreries were not even functional, "sometimes they [stars and planets] wouldn't come on, sometimes they wouldn't go off, sometimes they had holes in them, and mostly they didn't seem to be good likenesses." Not amused by the orrery show, the young Dickens lamented it would have been better "never to have been born" if this was the birthday treat.[42]

Another example of reminiscences of show visiting came from the English writer Edmund Gosse. Gosse's famous memoir *Father and Son* (1907) portrays a struggle between two characters as well as between generations. Gosse's naturalist father, Philip Henry Gosse, is depicted in the memoir as a stern, repressive parent. Gosse emphasizes the uneasy relationship between him and his father; his gradual coming of age and rejection of his father's fundamentalist religion are other main themes in the memoir.[43] Gosse recalls a minor event in his eighth year: he and his father paid a long anticipated visit to Mr.

Wyld's "Great Globe" in Leicester Square, one of the rare occasions in his childhood that he was ever taken to a place of entertainment. According to Gosse's diary, the exact date they visited the Great Globe was September 10, 1857. Wyld's Great Globe was a colossal spectacle focusing on geographical education. The proprietor James Wyld extended his ambition to other related subjects including astronomy, as he elaborated on the solar system in the compendium prepared to accompany the visitors.[44] The exhibition inside the Globe, however, disappointed Gosse. This small event in Gosse's childhood was unusual. Gosse describes his parents' indifference to entertainment: "Notwithstanding all our study of natural history, I was never introduced to live wild beasts at the Zoo, nor to dead ones at the British Museum. I can understand better why we never visited a picture-gallery or a concert-room."[45] A serious and religious man of science like Philip Henry Gosse would even consider Wyld's Great Globe in the heart of the West End worth a visit for his son. As in the case of Hollingshead's severe schoolmaster guardian, this surprising small event suggests that commercial exhibitions or lectures combining amusement and instruction, like the Great Globe and the orrery exhibition in theaters, were an important part of Victorian science culture and education. They were widely regarded as worthwhile attractions by contemporary parents and teachers.

A Fashionable Motivation for Science

Spectators attended Lenten astronomical lectures in theaters for various reasons. Hollingshead and his schoolmaster guardian went to the show expecting a performance in which "amusement and instruction combined." While the adult thought the balance between amusement and instruction ought to be tipped in favor of the latter, the child was more concerned about whether the show was fun and engrossing. Other audiences, mostly country people, were perhaps indifferent to astronomy, and Hollingshead speculated that they "probably thought they were seeing an ordinary play." The discontented sailor, too, drifted into the theater in search of a sword-combat play for entertainment. The motivations for people to attend a scientific lecture were no less diverse than the assorted audience composition.

It is common to presume that people went to a Lenten astronomy lecture because of the implicit religious messages it contained, or that people subscribed to the Royal Institution's courses of lectures because they wanted to obtain useful knowledge of scientific novelties. However, such presumptions are too simple and likely untenable. Some spectators, no doubt, went to theaters or the Royal Institution for the above reasons, but it is incorrect to think that the whole audience was motivated by the same expectations. As Jonathan Topham has shown regarding the readership of the Bridgewater Treatises, these titles were read in a variety of contexts, in which they "served radically different purposes and possessed radically different meanings."[46] Topham's observation agrees with reader-response and reception theory: once a book has left its production context, it is transmitted to multiple reading contexts in different cultural and social spaces. Topham particularly cites the example of the Bridgewater Treatise volume by William Buckland on geology and mineralogy: it was read by Sir Charles Bunbury as an illustrated guide to the paleontological collections of the British Museum. However, for the atheist Charles Southwell, that very volume served as a sourcebook for studying transmutation.[47] James Secord's study on eighteen-year-old Thomas Archer Hirst's reading of *Vestiges* also shows this diversity. Hirst did not originally read this infamous book for its discourse on evolutionary knowledge. He read it because of a debate with his close friend on whether science could enhance or undercut divine mystery and men's reverence for the Creator.[48]

The above studies of nineteenth-century readers of scientific publications remind us that attending scientific lectures, like reading, could have a variety of motivations. Fashion was noteworthy among numerous reasons motivating people to take part in a scientific lecture. Popular science, along with clothes, fine arts, and other curiosities in town were elements of fashion in nineteenth-century culture. Fashion might be vague and unpredictable, but it can be investigated in the spread of a particular commodity or a frequently mentioned keyword in contemporary diaries, letters, and literature. Books, exhibitions, and lectures associated with useful knowledge or scientific curiosities were precisely the commodities reflecting the fashion of science in the market. If a popular satirical publication like *Punch* lampooned or alluded to something, then the target of this ridicule must be famous or at least familiar to contemporaries. In the very first volume of *Punch*, the fashion of popular science and astronomy was a source of jokes and ridicule. The author imitates the style of "familiar science" to crack jokes related to celestial bodies from the Sun, Moon, and Earth to comets. "Our opinion is, that science cannot be too familiarly dealt with," the author remarked at the start of the article, "we are only following the fashion of the day, in rendering science somewhat contemptible, by the strange liberties that publishers of *Penny Cyclopædias*, three-half penny *Informations*, and two penny *Stores of Knowledge*, are prone to take with it."[49] This opening remark not only made fun of astronomical science itself but also mocked a mass of cheap publications advocating familiar knowledge.

The social culture of Victorian Britain undoubtedly contributed to the initiation of a scientific fashion. James Secord attributes the soaring popularity of *Vestiges* within a few weeks of its publication to the influence of the fashionable elite. This fashionable society, consisting of the aristocracy and the urban gentry, initiated intellectual and artistic fashion through social occasions at their mansions, townhouses, and clubs. New books, prints, and artworks were often displayed at various social events such as soirées, conversaziones, balls, and dinner parties. These novelties on display became conversation pieces for the participants in the events; exchanges of opinions and interests also took place in conversation. For authors and publishers, such social occasions were invisible but critical battlefields, deciding the sales and reputation of the works by word of mouth. The newly published *Vestiges* seized the initiative by successfully serving as a talking point at fashionable gatherings.[50]

Scientific topics, such as general knowledge about a particular field, novel invention, or discovery as well as books and periodicals circulating such information could serve as fashionable conversation pieces. The more formal talks on specialized topics between men of science took place at institutional meetings. Learned societies and institutions also organized conversaziones, meetings that planned for an invited speaker and the audience to engage in a discussion on prearranged topics. Soirées and salons were everyday occurrences where gentlemen and ladies participated in relaxed, informal conversations. Scientific topics were adequate options for rational small talk at these social events. Conversational etiquette, tacitly regulating what topic and language were appropriate on a particular occasion, developed through this polite culture.[51] An accomplished celebrity who attracted the crowd's attention at a party would be the "lion." In contrast, the "bore" with uncivilized manners and dull conversation would be unwelcome at a party.[52]

A satirical essay "A Discourse of Bores," written by John Poole and published in the *New Monthly Magazine* in 1838, reflected this culture of conversation and the attitude toward certain inappropriate manners.[53] Poole discussed the meaning of the word "bore," which he defined as "somebody who doesn't know when it is time for him to leave off doing something."[54] He then gave examples of different types of bores, derived from his acute observations of people on social occasions. One type was the "Superficial Bore," who is "constantly thrusting into your face his little farthing candle of knowledge, which sheds just light sufficient to render visible his own ignorance." Poole describes a figure named Sam Smatter, possibly an anonym or a synthesis of several real people, as the specimen of the Superficial Bore:

> He talks oracularly about hydrogen and oxygen; nitrates, muriates, carbonates, and sulphates; the gases, the acids, and the alkalis: though not always, perhaps, applying the terms with an exactness that would satisfy a Faraday. He disagrees with many philosophers concerning the nature of the electric fluid, and has made up his mind as to the true cause of magnetic attraction. And why should he not? or he has studied Pinnock's "Chemical Catechism" to very little purpose. Having qualified himself in astronomy by an attendance at two lectures on a Transparent Orrery, he bandies the sun, moon, and stars, as if they were so many cricket-balls; is not quite satisfied with the received theory of the tides; regrets that he is compelled to differ from Newton concerning the principle of gravitation; and . . . he sees clearly the practicability of catching a comet, provided he could but find the means of putting a little salt upon its tail.[55]

The Superficial Bore was a new class of spoilsport in the age of mass printing and cheap knowledge and a by-product of the flourishing of popular science. Poole satirized Sam Smatter's chief source of information, the *Penny Magazine*, from which he crammed, and the great portion of his talk throughout the week will be of the topics "according to the contents of his last number." This criticism ridiculed the Superficial Bore's shallowness and constant behavior of showing off. The satire on the Superficial Bore reflected a fashion of conversation about science. When this fashion became too extreme, it had a counterproductive effect: a poor imitation of the "lion" only produced the "bore."

The root of this polite culture of conversation can be traced back to the eighteenth century, when Enlightenment intellectuals advocated rational amusement and associated science with pleasing conversation and moral uplifting. The Irish writer Richard Steele's account in 1713 of the orrery, which was then still a novel instrument, was a representative description of this idea. Steele foresaw the educational as well as entertaining value of the orrery as a tool of enlightenment: "It [the orrery] administers the Pleasure of Science to anyone, . . . All Persons, never so remotely employed from a learned way, might come into the Interest of Knowledge, and taste the Pleasure of it by this intelligible Method." Steele even claimed that each distinguished family should possess an orrery "as necessary as . . . a Clock," for this machine would "raise a pleasing, an obvious, a useful, and an elegant Conversation."[56]

The fashion of rational amusement was at its zenith during the Regency and early Victorian eras. Richard Altick indicates that the prevalent rhetoric and practice of rational amusement in eighteenth- and nineteenth-century Britain was an acceptable and efficacious blend between what was thought to be "useful" or "improving" and the indulgence of human nature—enjoyment. It was a moral philosophy recognizing the human appetite for enjoyment yet optimistically considering that instruction harmonizes and utilizes human nature. In combining the two, however, instruction always took precedence over amusement. As Altick indicates, if a choice had to be made, "there was no question that instruction commanded a high priority over 'mere' enjoyment, of whatever sort."[57] Anne Secord's study of popular botany in the early nineteenth century also shows the significance of pleasure: the popularizers acknowledged that the promotion of knowledge would "best be served by adapting to the popular forms of pleasure of the audiences they wish to reach."[58]

Similar phenomena of fashion also appeared in the rapidly changing market of shows and exhibitions, and sometimes more avidly. Unlike intellectual fashion initiated by the genteel society, the fashion of shows tended toward the taste of a wider public but also spread through advertisements and word of mouth on social occasions. The exhibitions at the Egyptian Hall, for instance, were often sensational enough to convince thousands of people to flock to Piccadilly. *Punch* lampooned the fashion of seeing freaks and curiosities at the Egyptian Hall, calling it "Deformito-mania." The *Punch* satire claims that the public's taste for the "Monstrous" seems to have reached its climax; the walls of the Egyptian Hall in Piccadilly are placarded from top to bottom with "bills announcing the exhibition of some frightful object within." The *Punch* piece stated that the building itself would therefore soon be known as "the Hall of Ugliness."[59] Although these shows and spectacles might cater to the general public's vulgar taste and curiosity, some also advertised a more serious educational value. Thus, as we have seen above, adults like Gosse's father and Hollingshead's schoolmaster friend would bring juveniles for an instructive visit. The "Superficial Bore" Sam Smatter, who was overly eager to show off his erudition and was criticized by Poole, also absorbed information from astronomical lectures on the transparent orrery.

Astronomy was doubtless a significant subject in the fashion of popular science. Allan Chapman cites a newspaper clipping titled "The Astronomical Mania" to describe the crop of popular astronomical lectures in London during Lenten seasons.[60] The newspaper article, written by an anonymous journalist, reviewed the activities of eight lecturers, including James Howell and Dionysius Lardner, in London theaters during the previous Lent. Like the "Deformito-mania" proclaimed in *Punch*, the "Astronomical Mania" distinctly satirized the fashion of this kind of Lenten alternative to regular plays. A Lenten astronomical lecture, especially a famous one like Walker's at an elegant venue, was more a fashionable social event than a serious scientific meeting. Receiving knowledgeable instruction and sublime inspiration might be important justifications according to the producers and scientific or religious authorities, but participating in a fashionable show was a more effective motivation for ordinary people. This fashion was not confined to affluent audiences. The working classes were also willing to pay 1 shilling to gain admittance when C. H. Adams lectured at the Italian Opera House (Her Majesty's Theatre), although Chapman cites a journalist's suggestion that their real motive was "to avail themselves of the opulent delights of being in a great theatre for such a small sum."[61]

Another account demonstrating this astronomy fever is a long drollery poem "Love and Lunacy," written by the poet Thomas Hood and published in the *Comic Annual* in 1836, in which he makes a remarkable caricature of a lecture-going enthusiast.[62] The poem, according to Hood's biography sixty years later, was inspired by a joke with a friend, Lieutenant de Franck, while Hood was staying in Koblenz, Germany.[63] It is a farcical romance in which the protagonist Lorenzo, a fictional young man who loves astronomy, is keen to chase after scientific knowledge. The poet described Lorenzo:

> Ah! had he been less versed in scientifics,
> More ignorant, in short, of what is what;
> He ne'er had flared up in such calorifics;
> But he *would* seek societies, and trot
> To Clubs—Mechanics' Institutes—and got
> With Birkbeck—Bartley—Combe—George Robins—Rennie,
> And other lecturing men. And had he not
> That work, of weekly parts, which sells so many,
> The Copper-bottomed Magazine—or "Penny?"
>
> But, of all learned pools whereon, or in,
> Men dive like dabchicks, or like swallows skim,
> Some hardly damp'd, some wetted to the skin,
> Some drown'd like pigs when they attempt to swim,
> Astronomy was most Lorenzo's whim,
> ('Tis studied by a Prince amongst the Burmans);
> He loved those heavenly bodies, which, the Hymn
> Of Addison declares, preach solemn sermons,
> While waltzing on their pivots like young Germans.[64]

The boy Lorenzo, to some extent, is a character like Sam Smatter. Although Lorenzo is not as ostentatious as the Superficial Bore satirized by Poole, he is naively arrogant when talking about astronomy. In the plot, Lorenzo is in agony because of his girlfriend's ignorance about astronomy. He even blames himself for taking her to Covent Garden rather than to Bartley's orrery.[65] "Love and Lunacy" humorously presents an image of a young fashionable know-it-all, who is a lover of scientific knowledge derived from popular lectures and magazines, but who also easily quibbles about things that might be seen as obscure or irrelevant to other people. It was fashionable to read popular periodicals like the *Penny Magazine*, as well as to attend lectures at learned societies and mechanics' institutes. Lecturing men and working astronomers were important sources of knowledge through their lectures and

Passion Week at the Play

FIGURE 6.5. "Passion Week at the Play," *Cornhill Magazine*, vol. 5 (January 1862), 87. This cartoon humorously depicts the audience's reactions to Lenten amusements in a theater. © Chronicle / Alamy Stock Photo.

publications. Bartley's orrery was on a par with the work of the Astronomer Royal George Airy and other elite astronomers.

Polarized Truths

The disagreement of opinions among the audience of the astronomical lecture in Hollingshead's "At the Play" is evident. Hollingshead's guardian regarded the lecture as beneficial instruction and mourned its accidental interruption; Hollingshead, as a youngster, felt that the show was dull and the lecturer was patronizing and not making clear explanations. Other people in the audience were probably unsatisfied, too, according to their "bewildered and unhappy" faces, not to mention the sailor's disgruntled outburst. It seems that most of the audience in Hollingshead's story was not pleased with the show. No specific newspaper account of this lecture appeared, although sometimes newspapers mentioned an event only in passing. It is more likely that the show was just not newsworthy enough to elicit a mention. The overall reports of Lenten astronomy lectures in contemporary periodicals usually focused on a few celebrated performances, such as those of Bartley or C. H. Adams. It is unclear how numerous other lesser-known Passion Week astronomical amusements were received.

One of the illustrations in Hollingshead's article, titled "Passion Week at the Play," depicts the confused or disappointed faces of the audience (fig. 6.5).[66] This illustration neither directly connects with Hollingshead's story nor indicates a particular Lenten amusement scene. It is nevertheless a comical sketch of the audience's response to a Lenten amusement. Since heavenly bodies usually replaced earthly dramas in theaters during Passion Week, this illustration could be a lampoon of Lenten astronomy shows. Besides, given that this illustration was inserted into an article on the experience of visiting a Lenten astronomy lecture, it is hard not to think of the connection between the two. Unlike the positive claims of advertisements or newspapers, it seems that these spectators were not amused at all. When reviewing amusements in provincial towns or the country a *Saturday Review* journalist straightforwardly criticized local amusements, especially those "informative" lectures of all sorts that were "more or less stupid and dreary."[67]

The composition and motivations of the audience were diverse, as were their reactions. Reader-response and reception theory suggest that the meaning of a text changes. Readers and audiences make their own interpretations depending on personal experience, background, and beliefs, rather than passively accepting the original messages the author has tried to transmit. It is never easy to reach an overwhelming consensus on a matter. Thus, a varied reception from an audience is nothing strange. Astronomy can be involved in controversy; an astronomical lecture can win one spectator's approval but simultaneously incur the wrath of another. Such dissension is especially apparent when a lecturer presents controversial issues related to religion and politics in a high-profile way.

In this book, I have shown the profound connection between astronomy and religion, as well as the controversial issues in which astronomy associated: the science of progress—the nebular hypothesis—and the plurality of worlds. We have seen the examples of J. P. Nichol's radical initiative and John Herschel's reserved stand on the nebular hypothesis debate. As the evolutionary debate later in the mid-nineteenth century showed, the tension during this period was not merely on an intellectual or theological basis. These debates are better understood in the social context of conflicting ideas of conservatism and radicalism.[68] The complex entanglements between politics, religion, and science, unsurprisingly, are reflected in the reception of popular astronomy lectures. The following example is an account of a scathing attack on an astronomical lecturer by an atheist spectator. Their disagreement also extended beyond religion to politics.

Published in the radical journal the *Republican* in 1825, this is a letter from a "Lover of Truth" to an itinerant lecturer, Mr. Rogers. The letter was dispatched from Portsmouth and dated January 6, 1825. It strongly rebutted Mr. Rogers's lecture of the previous evening, especially his criticism of the "Infidels." One of the infidels reproachfully pointed out by Mr. Rogers was Richard Carlile, so the Lover of Truth took the liberty of sending Mr. Rogers three issues of the *Republican*, of which Carlile was the editor. The Lover of Truth attacked the religion advocated by the lecturer, declaring his belief in atheism, and expressed complete contempt for the lecturer.[69] The lecturer, according to the Lover of Truth, "did not exhibit one single rational and valid argument, to prove the existence of such a God." The Lover of Truth not only challenged the religion of Mr. Rogers but also called into question his qualification for lecturing on astronomy:

> As to your descriptions of the Moon, Tides, &c. they were upon the whole very poor, and far beneath what I expected to hear from you; and only proved to me, that you have not been a deep reflecting Astronomer. Although, before last night, I never saw an Orrery, or ever heard an astronomical or even lunary lecture, I have the vanity to think . . . that were I only in possession of an equally good voice, power of articulation and delivery as yourself, and had an equal knowledge of the meaning of words, I

could prove myself a much better Selenographer than you at present are. ... I would endeavour to become an astronomical lecturer, and your itinerant opponent; and assure you that I should not despair of being able to drive you out of the field, in a very few months; unless you should cease to pursue your present Theistical jargon, and in place of it begin to develop a more rational Theory, and correct knowledge of the attributes of all-puissant Matter; and also more clearly shew the relations and conditions of the stupendous glomerated bodies that float in the universe.[70]

In addition to astronomical subjects such as the Moon, the tides, and the orrery, Mr. Rogers also talked about the microscope and cited its vast power "with a view to confound and refute Infidels." To fight back, the Lover of Truth criticized the lecturer's reasoning and conclusions as fallacious in that the atheist saw the power of the microscope in the contrary view, claiming that the discoveries revealed by the instrument could only turn people into atheists: "It is those, who reflect and range fearlessly from the Minutest point, to the Greatest extent of space, that their imagination can possibly carry them; or from the Greatest to the Smallest conceivable atom of Matter or point of Space, who become Atheists; for these things shew them that there is no room left for a God to exist in. But if you should say, that God is not Matter, then, it is only a NONENTITY that you prate about."[71] The Lover of Truth originally wanted to question Mr. Rogers at the lecture, but had given up this provocation, since there were "interesting parties" in his way. It implies that the lecturer or the host had prepared some sort of security measures to prevent possible confrontation. The Lover of Truth denounced such a scene and concluded with the subversive view that "every honest man will boldly spurn at every Law and Custom that has its foundation in tyranny and ignorance."[72]

This protest letter is an unusual account of a spectator who openly opposed an itinerant astronomical lecturer. Very little is known about either the dissenter or the lecturer. The lecturer was probably A. F. Rogers, who had presented discourses on astronomy at the Masonic Hall, Bath, during Lent in 1824. Rogers's lecture had a specific religious agenda—refuting objections that the "Infidels" leveled against "Sacred Writings," as he advertised in the poster.[73] In the letter, the Lover of Truth claimed to be "a poor unlettered, but rational and reflecting mechanic." The periodical in which the letter was published, the *Re-*

publican, was a radical journal edited by Richard Carlile, a famous activist and declared materialist in early nineteenth-century England.[74] Carlile had been involved in many political activities and on the editorial staff of several radical newspapers in London since the 1810s. His efforts to popularize the works of Thomas Paine, plus the ideas of radicalism and atheism he disseminated through his newspapers, had provoked both the government and the Church against him. He was a witness to the Peterloo massacre and published firsthand accounts in his newspapers, which led to further charges of seditious libel and a six-year sentence between 1819 and 1825. However, while Carlile was imprisoned, he enjoyed a certain extent of liberty to write and he attempted to smuggle his manuscripts out of prison. He thus continued writing for the *Republican*, and the publication of the journal was maintained by his wife and supporters. A monograph, *An Address to Men of Science* (1821), in which Carlile urged men of science to "stand forward and vindicate the truth from the foul grasp and persecution of superstition," was published under the circumstances.[75] Nevertheless, Carlile himself could not have attended Rogers's lecture in January 1825, because he was not released from the prison until November of that year. The Lover of Truth must have been an acquaintance or a keen supporter of Carlile.

Even though Carlile was then in jail, he certainly knew about the entire dispute between the Lover of Truth and Rogers. Carlile's article "Astronomers and Astrologers" directly followed the letter of the Lover of Truth in the same issue of the *Republican*. This article was a more sophisticated attack on Rogers, in which Carlile claimed he had traced the itinerary of Rogers by "having had his bills sent to me, from Stockport to Norwich, and now round to Portsmouth." The Lover of Truth might be a "scout" sent by Carlile to spy on Rogers. Carlile taunted that since the theme of Rogers's lectures was "to refute infidelity," he welcomed Rogers to lecture to him, and "we will then soon see if astronomy demonstrates a God." Carlile also belittled his opponent's motivation, asserting that Rogers was merely a profit-chasing demagogue, who only wanted to "fill his pocket, or his belly, and cares not how."[76]

Carlile openly voiced his antipathy toward religious elements in science. He was a determined champion of materialism, and his religious position progressed from deism to atheism during his life. Carlile supported scientific education promoted by contemporary reformers like Henry Brougham, but he went further: Carlile asserted that religion should be removed from education. In *An Address to Men of Science*, Carlile particularly criticized men of chemistry and astronomy, for they "have openly countenanced systems of error and imposture, because the institutions of the country were connected with them; or, because they feared to offend those persons who might be deriving an ill-gotten profit from them."[77] He denounced the astronomers who overtly support Christian dogmas as an "astrologer," and claimed that such a theistical astronomer is "a corrupt and wicked hypocrite, and a disgrace to the science which he studies, practises, or teaches."[78]

The quarrel between Rogers and Carlile or the Lover of Truth demonstrates a radical attitude outside the mainstream of widespread religious narratives of popular astronomy. Undoubtedly, religious narratives, especially those related to natural theology, profoundly influenced early nineteenth-century popular astronomy. Christian people and institutions, whether Anglican or evangelical churches, contributed much to the dissemination of astronomical knowledge and the inspiration of the general public's interest in this subject. From Lenten lectures to the Bridgewater Treatises, a religious tone was very common in astronomical popularization during the first half of the nineteenth century. While devout Christians employed astronomy as an important instrument of moral and religious teachings, atheists and radicals had an antipathy to religious interpretations. The writings of Richard Carlile were representative examples. Opponents of the establishment used astronomy to support their radical agenda, too. Carlile's followers believed that science only leads to atheism. The radical astronomer J. P. Nichol also promoted the nebular hypothesis to endorse the science of progress.

Nevertheless, such a severe attack from the atheist side on a religious astronomy lecturer still seems uncommon. Bartley and Walker, two of the most famous contemporaries in this trade who also extolled the sublimity of the Creator, had never attracted a similarly fierce attack. In fact, Carlile did not totally oppose what they were doing. He was "much pleased to see that a number of gentlemen are giving lectures on Astronomy in all our towns and cities of any note." Carlile reasoned: "Such men [astronomical lecturers] are worthy of support in preference to the Priest, and although they may jointly, from fear, or other motives, attempt to mix up religious dogmas with their scientific lectures, I know that it must tend to a due enlightenment of the public mind. An Eidouranion or Orrery to have been displayed a few centuries ago would have gathered a pile of faggots for the lecturer, and he would have been burnt as a darling blasphemer, and his machine with him, as the devil's workmanship."[79] Carlile did not want to undermine the benefit of the enlightenment brought by astronomical lectures, despite the mixture of religious elements within them. However, Rogers's overt theme of "refuting Infidels" and his criticism of Carlile's works ignited the fuse of this dispute. The letter by the Lover of Truth started a war that broke out when an astronomical lecturer crossed the line and tangled with a fervent opponent.

The case of Carlile and the Lover of Truth was not the only religious dispute inspired by popular astronomy lectures. In chapter 2, I briefly mentioned a would-be lecturer who attempted to promote a geocentric model and thus refute Newtonian cosmology. This anti-Newtonian lecturer was probably a follower of Muggletonianism, an obscure and small sect of dissenting Protestants. The tract *Two Systems of Astronomy* (1846) by Isaac Frost, with many beautiful illustrations comparing geocentric and Newtonian models, was a representative work elaborating the universe according to Muggletonian cosmology. Since its establishment in the mid-sixteenth century, Muggletonians had always been a minority and had maintained a community relatively closed to outsiders. They were not outspoken of their faith, nor did they seek to convert nonbelievers.[80] So why did Muggletonians slightly change their attitudes in the early nineteenth century and even publicly present works like *Two Systems of Astronomy*? Francis Reid argues that the popularity of astronomical lectures at the time provoked the Muggletonians' response. Muggletonianism was a plebian movement in origin, so its followers were mostly artisans or people from the lower social classes. Astronomical lectures aimed at a mass audience or held in a working-class venue such as a mechanics' institute caused such lectures to become more easily accessible to Muggletonians than ever before. The subjects of the Newtonian system and the plurality of worlds, which were often elaborated in contemporary astronomical lectures, conflicted with the Muggletonians' understanding of Scripture. Moreover, many of the lecturers who presented such cosmological views were liberal Christian supporters of natural theology or evangelical clergymen like Thomas Chalmers, whose "heresy" was intolerable to Muggletonians.[81] Therefore, actions or works like *Two Systems of Astronomy* defending Muggletonian cosmology emerged in the early nineteenth century. Despite the very opposite stand held by Muggletonians to the atheist Carlile and the Lover of Truth, they all reacted to popular astronomy lectures at the time.

Although the debate in the Carlile–Rogers dispute was over theism and materialism, considering Carlile's political radicalism and the agenda of the *Republican*, this conflict was also political. Carlile was the figurehead of a potential revolution through his influence on the working classes in this period. The Lover of Truth asserted that people would boldly "spurn at every Law and Custom that has its foundation in tyranny and ignorance" in the conclusion of his letter. The argument could be stirred up by a mixture of religious skepticism and political radicalism, though we will never know if Rogers had an obvious political agenda besides his religious belief. Unfortunately, there are no accounts of Rogers or the subsequent development of the dispute. Nevertheless, Rogers's agenda on "refuting Infidels" clearly expressed his views about God's truth, as many other contemporary Christian publications asserted. For example, an article on astronomy in *Youth Magazine*, an evangelical periodical aimed at Christian youth, claimed: "Every reader of this paper might be enabled with sincerity and truth to adopt this language as his own, when surveying the wonders of the heavens."[82]

Intriguingly, both atheists and Christians used the word "truth" in their rhetoric. Whether God's truth or atoms' truth, each side used science to endorse their belief. A piece of scientific achievement or discovery could lead a devoted Christian and a subversive radical to reach opposite conclusions. Lord Rosse's giant telescope at Parsonstown, for example, inspired J. P. Nichol to give his lecture audience a vision of an economic and political future, in which the prospects of the working classes would be demonstrated by experts as clearly as the heavens revealed by Rosse's great instrument. In contrast, Nichol's opponent Thomas Romney Robinson, an Anglican priest and the longtime director of the Armagh Observatory, told his lecture audience that the giant telescope would reveal "God is there."[83] The giant telescope could lead lecturers to very different views of cosmology, politics, and morality. Likewise, the power of the microscope was another common metaphor used by lecturers and could lead to different views about materialism, natural theology, and the plurality of worlds.

It is always a challenging task for researchers to chart a comprehensive map of the audience or readership for a scientific work. In this chapter, I have demonstrated various methods and clues that could indicate an outline of the audience of popular astronomy lectures in the nineteenth century. From the upper classes to the working classes, from adult guardians to juvenile pupils, from fashionable socialites to revolutionary radicals, the wide spectrum of audiences for astronomy lectures was as remarkable as the diversity of venues where they were held. Among the various groups of spectators, women, teenagers, and parents were particularly important target audiences, especially for Lenten astronomy lectures. This tendency accorded with the conventional emphasis on the educational value of astronomy since Enlightenment polite science culture. It also reflected a majority consensus that astronomy, with an appropriate accommodation for moral and religious inspiration, was a suitable and beneficial science.

Nevertheless, the motivations and responses of audiences varied: rather than attending astronomical lectures for scientific reasons, contemporaries might attend for social reasons, such as following fashion, collecting conversation pieces, or simply seeking amusement. Audiences could also disagree with the lecturer's agenda or performance. Sometimes the disagreements were intense due to contrary religious or political views. The Carlile–Rogers dispute best exemplifies the conflicts of faith and tension between supporters of different ideologies in nineteenth-century popular astronomy. It also reflects the fact that science in the nineteenth century was a contested space. Lecturers and audiences all had their own say.

EPILOGUE

An Enduring Legacy

The front page of the London-based journal the *Athenaeum* was like a clearing house for distributing commercial intelligence. Advertisements and notices related to science, literature, education, and fine arts usually occupied the front page of each issue. An advertisement appeared in July 1865 addressed specifically to lecturers, institutions, and "persons seeking a profession." The message that the advertisement carried, in fact, implied the end of an era: "For sale, the original eidouranion, or large transparent Orrery, with which the late Mr. Deane, Franklin-Walker [sic] Lecturer on Natural and Experimental Philosophy, illustrated the successful and popular Lectures on Astronomy which he gave in London for many Seasons during Lent. Also, the extensive philosophical apparatus lectured on by him at the Public Schools and Colleges."[1] The sale, however, seemed to attract no buyers and the same advertisement reappeared four months later.[2] The obituary of Deane Franklin Walker in the *Gentleman's Magazine* shared the same sentiment. In it the journalist remarked that the world and science "have moved on mightily" since the time when Walker's lectures and the starry scenery of the eidouranion "were in the height of their popularity," and "the generation that used to talk of him has passed away."[3]

The death of D. F. Walker and the unsuccessful sale of his philosophical property embodied the decline of a trade. The Walkers had been the epitome of public philosophical lectures and the theatrical performance using the transparent orrery since Adam Walker's lecturing business in the late eighteenth century. No known family mem-

EPILOGUE

bers of the Walkers inherited this legacy after D. F. Walker. At the moment when the late D. F. Walker's philosophical property was for sale, many other once-celebrated lecturers also faded from the stage. C. H. Adams had already retired from his more than thirty years of annual lecturing, and George Bachhoffner was no longer the manager of the Colosseum. This does not mean that no Lenten astronomical lectures took place thereafter, but this seasonal amusement became less and less a focus of public attention.

There had been attempts to revive Lenten astronomical lectures, but these efforts were fruitless. The renowned science showman John Henry Pepper, for example, once tried to restore astronomical lectures at the Royal Polytechnic Institution after Bachhoffner's departure. Advertisements for Pepper's lectures on astronomy had appeared in newspapers intermittently since the early 1860s.[4] During the Lenten season in 1871, the *Theatrical Journal* reported Pepper's astronomical lectures on Thursday afternoons and revisited the faded convention of Lenten stage astronomy. "Many of us must remember the time when the poor player was not allowed to exercise his calling during the Wednesdays and Fridays of Lent, his place on the boards being taken up on these days by mountebanks, negro melodists, and the like," the journalist pointed out, "the best of the substituted exhibitions being Mr. Adams's annual Orrery at the old Adelphi." The reporter described Professor Pepper's astronomical lecture as a reproduction of the "once popular representation of the revolutions of the heavenly bodies."[5] In 1872, however, Pepper resigned from the Royal Polytechnic due to disagreements with the board over the extent of his autonomy. He then performed his scientific shows at the Egyptian Hall and went on tour in the United States, Canada, and Australia.[6] It is unclear whether he lectured on astronomy after he left the Polytechnic.

This *Theatrical Journal* article shows that the ban on regular plays during Lent was no longer enforced in 1871. Before the mid-nineteenth century, Lenten restrictions on dramatic performances in Britain had already been flouted by many minor theaters. By the 1860s, the restrictions seemed completely lifted as the public no longer refrained from theatergoing and pleasure seeking. As a result, the initiative for astronomical lectures as an alternative to regular plays during Lent was gone. Besides, conventional Lenten astronomy lectures faced more competition with the arrival of new entertainment venues, such as music halls, which provided a wider range of variety shows and grew rapidly after the midcentury. As choices of amusements increased, Lenten astronomical lectures probably lost their entertainment as well as educational appeal to audiences, owing to their out-of-date performance style and commonly shared religious tone.

The environment of London entertainments had changed drastically in the mid-Victorian era, particularly after the Great Exhibition of 1851. Along with many other types of conventional shows including panoramas and other exhibitions, Lenten astronomy lectures were also on the wane. Richard Altick argues that several factors changed the pattern of city life and leisure, hence contributing to the decline of certain conventional shows in what he calls the "post–Crystal Palace era." These factors included new means of transportation and the changing status of social fashion.[7] Railway infrastructure expanded rapidly in England between the 1840s and the 1850s. Broader railway networks allowed people a convenient way to travel to rural regions, which changed patterns of leisure and partly undermined the West End's status as a recreation center.[8] The public's taste and expectations for the standards of exhibition also changed. Conventional shows encountered increasing pressure not only because of changing fashions for entertainments but also because of institutionalized science and new types of educational establishments. The Great Exhibition of 1851 spoiled spectators on an incomparable scale with good value on one-shilling days, which made any other Leicester Square spectacles seem meager. Individually operated lectures and exhibitions now faced more severe challenges from larger-scale sites such as the relocated Crystal Palace at Sydenham, and later from state-funded institutions, museums, and galleries on the South Kensington model.[9]

The demise of Lenten astronomical lectures and old-school entrepreneurs did not mean the end of private astronomy lecturing. Independent astronomical popularizers, such as Richard Anthony Proctor (1837–88), Arthur Cowper Ranyard (1845–94), and Agnes Mary Clarke (1842–1907), remained active in writing or lecturing in the last three decades of the nineteenth century. However, their connections to an institutionalized science network were much stronger than the last generation of private entrepreneurs. Both Proctor and Ranyard were university graduates and fellows of the Royal Astronomical Society. They first established their credentials as practicing astronomers and later earned their livings as editors, popular authors, and lecturers. Proctor served as the honorary secretary of the RAS and was the editor of its organ *Monthly Notices of the Royal Astronomical Society*. In addition to his institutional service, Proctor also founded the journal *Knowledge*, a popular science monthly with extensive astronomy coverage. He also did lecture tours across the Atlantic.[10] Unlike Proctor, who was prone to clash with other members of the RAS, Ranyard enjoyed a better relationship with colleagues and served as a longtime member of the RAS council and as its secretary. After Proctor's sudden death, Ranyard took over Proctor's editorship of *Knowledge* and completed unfinished books left behind by Proctor.[11] Clerke, as a preeminent example of a woman science writer, enjoyed friendly relations with practicing astronomers despite her lack of formal education. Scientific establishments later celebrated her contributions to popular writing: she received the Actonian Prize from the Royal Institution in 1893 and was elected an honorary member of the RAS in 1903.

In scientific institutions, astronomy was still a popular subject for lecturing. The Royal Institution, for example, maintained strong connections with the scientific elite and invited the Scottish astronomer Robert Grant to give courses in astronomy for three consecutive years from 1868 to 1870. Grant was the successor to J. P. Nichol's chair of astronomy at Glasgow University.[12] The brightest star in the firmament of British popular astronomy in the late nineteenth century, however, was the Irish astronomer Robert Stawell Ball. Ball was the Andrews Professor of Astronomy at Trinity College, Dublin, and later the Lowndean Professor of Astronomy at Cambridge. He was deeply involved in public lecturing and popular writing, thus earning a reputation as a skillful and charming popularizer. Ball's frequent lecturing at the Royal Institution, along with publishing bestsellers such as *The Story of the Heavens* (1886) and *Star-Land* (1889), marked his success in popularizing astronomy. He was also renowned for his adept use of lantern slides, a visual aid increasingly used by private astronomical showmen since the early nineteenth century.[13] Ball's great success in both academia and the popular market marked the beginning of a new era in which professional experts of science occupied the space of popular science left behind by "amateurs." Nonpractitioner popularizers of the old school, whether Enlightenment natural philosophy lectur-

EPILOGUE

ers like the Walker brothers or self-taught artisan entrepreneurs like John Bird, had lost their place in an increasingly institutionalized and specialized world of science. The scenes of public astronomy lecturing after the 1860s were quite different from the earlier marketplace.

The popular astronomy lecturing trade during the period between 1820 and 1860 as covered in this book changed significantly. At the beginning of the 1820s, a large number of private entrepreneurs were active in the marketplace of popular astronomy. Many of them inherited the conventional approach, subjects, and discourse of public philosophical lecturing in the eighteenth century, with a touch of the apparatus and showmanship of the new stage astronomy. The most successful of them, such as the Walkers and George Bartley, had a cultural and scientific reputation among the public that could rival the standing of scientific elite and institutional lecturers. However, by the end of this period, their status in the marketplace had been largely replaced by a new generation of institutionalized popularizers. Although the lecturing trade changed so much over time, popular astronomy persisted as a significant cultural form throughout the period and beyond. As Ian Inkster remarks, "Few of the sciences could survive as popular cultural forms in the face of the new rigor."[14] Astronomy's long popularity among the public is attributed to its diverse appeal, especially its content and implications in the areas of religion, morality, and politics, as we have seen in this book.

Mapping an Uncharted Terrain of Astronomy

This book has explored popular astronomy lecturing in Britain throughout the first half of the nineteenth century, covering its different aspects including practitioners, sites, curricula, apparatus, and audiences. The tools employed in this research have benefited greatly from scholarly works on historical studies of popular science and scientific publishing in the past few decades, especially the critical examination of a broader coverage of "text" that includes advertisements, pictures, maps, conversations, and all sorts of ephemera. These techniques have been developed and demonstrated in works such as James Secord's *Victorian Sensation* and Bernard Lightman's *Victorian Popularizers of Science*. I expect this book to serve as a supplement or "prequel" to the latter, which also focuses on nonelite popularizers of science but with a greater emphasis on the second half of the nineteenth

EPILOGUE

century. As I have shown, a wide spectrum of astronomical popularizers was active during the Regency and Victorian eras. Lecturers spoke about the heavens at various sites ranging from mechanics' institutes to London's West End theaters. The styles and languages of lectures were adapted to specific settings and audiences. A discourse delivered by a Cambridge professor at the Royal Institution on Friday evenings was quite different from the performance of a professional showman on the stage of the English Opera House during Passion Week.

Of diverse popularizers of astronomy, I specifically compare institutional men of science and private lecturing entrepreneurs, two presumably distinct groups. This dichotomy is certainly a rough division, despite its effectiveness for analyzing the affiliations of practitioners of popular astronomy. Therefore, the comparison should be made with care. The boundary between the two groups was blurred and porous in the first half of the nineteenth century; the identity of an astronomical lecturer was mixed and dynamic. The demarcation between "professional" and "amateur" or between "institutional" and "private" was slight, and sometimes a practitioner could belong to both.[15] Furthermore, popularity and credibility did not necessarily follow one's identity as a "professional" or "amateur" practitioner. When contemporary laypeople looked for knowledge of astronomical phenomena, they could seek, for example, George Bartley's orrery lecture, the Religious Tract Society's numerous publications, and William Whewell's Bridgewater Treatise. As Ruth Barton and Allan Chapman indicate, there was no consistent association of amateurs with lower scientific standing before and during the mid-Victorian era, especially in the realm of astronomy where most practitioners were amateurs on financial grounds.[16]

Although all the different classes of popularizers are covered in this book, the emphasis is on a group of private lecturing entrepreneurs, which included itinerant lecturers, spectacle exhibitors, and theater performers. Their astronomical expertise was often self-taught; they usually lacked connections with or had weak ties to scientific institutions and practicing astronomers. William and D. F. Walker, George Bartley, and C. H. Adams, were the most successful based on the longevity of their businesses, or the fact that their shows were widely mentioned in contemporary popular literature. Their ubiquitous errands across the metropolis and provinces were common scenes in early

EPILOGUE

nineteenth-century Britain. These private entrepreneurs were able to reach many audiences and were regarded as legitimate presenters of astronomical knowledge despite their lack of institutional affiliations. For lecturers who came from lower social classes, like John Bird and Robert Children, astronomy lecturing was even a potentially lucrative venture that afforded them better income and social status.

Some other astronomical lecturers, such as John Wallis and George Bachhoffner, had stronger institutional affiliations as their activities were more associated with scientific establishments like the Royal Institution and the Royal Polytechnic Institution. They coexisted and competed with private entrepreneurs in the marketplace. Sometimes they even became independent entrepreneurs themselves, as did Bachhoffner in his project to salvage the Colosseum in his later career. An institutional lecturer of astronomy, however, was not necessarily a scientific practitioner working on or supporting astronomical research. Although Wallis and Bachhoffner were employed by scientific institutions to deliver astronomical lectures, neither of them was involved in the communities of Grand Amateur astronomers or professional observatory staff. They were definitely outside the scientific elite of British astronomy, namely, the established "Greenwich-Cambridge axis" represented by Astronomer Royal George Airy and his circles.[17] So were private lecturing entrepreneurs who performed on theater stages or in assembly rooms of provincial towns.

Bernard Lightman illustrates the prominent place of nonpractitioner popularizers by analogy with a survey "mapping out the topography" of nineteenth-century British science. As the historiography of nineteenth-century British science was first dominated by scientific gentry and Oxbridge clergy, and later by scientific-naturalist reformers in the second half of the century, the voices and roles of nonelite outsiders have usually been neglected by mainstream scholars. Thus, Lightman poetically describes his reconfiguration as a means of allowing groups from outside the intellectual elite "to emerge from under the shadow cast by the mountainous elevations formerly occupied by scientific naturalism."[18] In agreement with Lightman's objective, my research also aims to secure a distinctive place for those who were frequently at odds with the scientific elite. As this book has shown, private entrepreneurs could have a widespread and major influence on contemporary audiences and popular culture, but they have often been forgotten by present-day researchers. The task of this book therefore has been to map an almost uncharted terrain of British astronomy.

A parallel topic discussed in this book is the "theatrical turn" of popular astronomy lecturing that began in the late eighteenth century, and had consolidated before 1820. Lenten astronomical shows were especially associated with this trend. Theater facilities helped achieve ideally dramatic visual and acoustic effects and thus enhanced astronomical lectures to a new level of amusement. Although the scientific elite did not regard Lenten astronomical shows or the like as seriously scientific tours de force, many contemporaries deemed these shows valuable in educational and recreational perspectives. The theatrical turn was not confined to private lecturing circles; institutional popularizers of astronomy could not neglect the trend, as similar techniques in using visual aids were increasingly being adopted. Robert Ball's efforts to incorporate lantern slides into his popular lectures best exemplified the trend that even men of science needed to compromise on the amusement side when dealing with lay audiences.

The Power of Emotion

The theatrical turn also helped the "sublime" representation of astronomy to gain currency in the wider audience. The distinctive feature of astronomy as the sublime science in both the discourse of its disciplinary knowledge and its image in popular culture has been discussed in this book. The astronomical sublime had never been a fresh concept for Victorian audiences; authors in the eighteenth century and earlier had written extensively on the theme of the sublime. Marveling at and admiring the starry sky and natural celestial phenomena were common long before. Not until the theatrical turn of astronomical lecturing applied theater facilities and techniques, however, were the wonders of the universe and the secrets behind them presented so vividly and entertainingly to a wide audience. The invention of the transparent orrery and the introduction of lantern slides in popular astronomy provided lecturers with powerful tools enabling them to present knowledge visually that not long before could only be conveyed using descriptive texts and a few diagrams. The rhetoric of natural theology about the Creation and heavenly order, which had originally been emotionally provocative, was now further empowered by the lecturer's skills in elocution and performance and the timely use of music.

EPILOGUE

Just as the astronomical sublime was meant to inspire feelings of surprise, awe, and wonder at the spectacles of the cosmic structure and revolution (and, for religious people, the Supreme Being behind these displays), the secret formula of popular astronomy is emotion, rather than rational understanding as many people would assert. This is not to say that rational and scientific understanding of the physical universe is irrelevant. The intellectual progress that human reason could achieve, such as the correct understanding of the universe from the Ptolemaic to the Copernican and the Newtonian systems, are subjects that astronomical popularizers since the eighteenth century have been happy to talk about. However, what really appeals to the audience is not the mere knowledge itself, but the prospect of satisfying their emotional needs, whether it is basic amusement for the senses or more sublimated contemplation on a religious or philosophical level. Jan Golinski has pointed out that the merit of the Walkers' eidouranion was its ability to elicit religious awe with the use of the sublime features to include some of the implicitly radical ideas of Enlightenment thinkers, such as the notion of extraterrestrial life on other planets. The Walkers had found a novel way to convey sublime experience to the audience, thereby "linking astronomy to the aesthetics of the sublime."[19] The "Ouranologia" presented by George Bartley, and many other similar Lenten astronomical lectures in the early nineteenth century, followed the same path initiated by the Walkers.

Alan Gross has suggested that the search for the sublime is a traditional essence in popular science throughout the history.[20] Gross traces sources that rendered the literary sublime in ancient times, then reflects on the natural sublime by Enlightenment or Romantic authors in the early modern age, and the transformation to what he calls the "scientific sublime" in modern popular science. In its greater focus on analyzing the traits of the sublime in popular science works by twentieth-century and present-day scientists from Richard Feynman to Richard Dawkins, Gross's survey of historical transitions of the idea of the sublime is slightly crude, but its insight into the relationship between popular science and the sublime is nevertheless noteworthy. In particular, Gross indicates that because science has become the cultural authority in secularized society, it replaces biblical narratives and theological dialectics as the new source for explaining fundamental questions such as the origin of life and the world. Therefore, he claims,

EPILOGUE

"science, not God, is the only wellspring of the sublime" in our present day.[21]

Rather than demonstrating the power of science itself, Gross's notion that science replaces God to become the new source of the sublime is more about revealing the power of emotion in popular science. Disciples of scientific naturalism such as Thomas H. Huxley and John Tyndall were dominant forces in late nineteenth-century popular science. They have taken over from their natural theologian opponents the mission of evoking the sublime.[22] On the other hand, whether they liked it or not, the new generation of professional scientists who wanted to speak to the public could not totally escape the language, narratives, and representation used by earlier popularizers. They had to appeal to both sense and sensibility, just as their natural theologian predecessors had done. Professional scientists might not appeal to religious reverence, but they still framed their popular work in the language of awe and wonder about the universe.

This makes popular astronomy lecturing in the nineteenth century, despite different scopes and apparatus, part of a continuous "tradition" that has endured up to today's popular science. All the techniques derived from the theatrical turn to display the astronomical sublime have a profound impact on popular astronomy that continues to the present. Today's audiences are familiar with those purpose-built dome theaters in museums and science centers, which used planetarium and film projectors to show the night sky or educational (often entertaining) movies. The invention and popularization of modern planetarium projectors—the first model was built by Carl Zeiss optical works in Jena, Germany, in 1923—is beyond the scope of this book.[23] Nevertheless, the legacy of the theatrical turn remains prevalent. Although the technology of modern planetarium projectors is new, the techniques required in such astronomical shows, from discourse and stage presence to the use of visual, vocal, and musical performance, had already emerged in London's West End theaters in the early nineteenth century.[24]

In chapter 5, I cited the Royal Institution lectures by renowned astronomer and scientific celebrity Carl Sagan as a symbol of continuity in popular astronomy. Although he was an outspoken champion of the rational values of science, Sagan proficiently used the expression and rhetoric of the sublime in his broadcasts and writing. He stated in *Cos-*

mos: "Today we have discovered a powerful and elegant way to understand the universe, a method called science ... science has found not only that the universe has a reeling and ecstatic grandeur, not only that it is accessible to human understanding, but also that we are, in a very real and profound sense, a part of that Cosmos, born from it, our fate deeply connected with it."[25] It is not difficult to find some traces of the sublime discourse of popular authors since the eighteenth century in Sagan's statement. For generations and generations, popularizers of astronomy have continued to evoke the awe and wonder of the universe.

APPENDIX

ASTRONOMICAL LECTURERS IN BRITAIN, 1800–1870

This list includes people known to have delivered at least one public lecture on astronomy in Britain between 1800 and 1870. It consists of lecturers of different backgrounds, affiliations, and agendas. Names in bold font indicate that person's presentation of multiple lectures in two consecutive years, or lectures in any three nonconsecutive years. Names preceded by an asterisk represent those who receive special attention in this book and are given relatively detailed biographical coverage.

Sources for this compilation include advertisements, playbills, syllabi, reviews, and news reports in periodicals. The main limitation of the list is its exclusion of lecturers who did not advertise or who received coverage outside these sources. These sources are given in the chapter endnotes and the bibliography, and thus not repeated here.

The format of each entry follows this order:
Name (birth and death) / affiliation or occupation / place of lecturing / years of lecturing or active period

*Adams, Charles Henry** (1803–71) / private; schoolmaster / King's Theatre (Her Majesty's Theatre); Haymarket Theatre; Lyceum; Adelphi Theatre; Princess's Theatre / 1830–61

APPENDIX

Airy, George Biddell (1801–92) / Astronomer Royal (1835–81); FRS; FRAS / Royal Institution / 1848, 1851, 1853–55

Babbage, Charles (1791–1871) / FRS / Royal Institution / 1815

***Bachhoffner, George Henry** (1810–79) / principal, Royal Polytechnic Institution (1838–55); manager and proprietor, Colosseum (1856–62); FCS / Polytechnic Institution; Colosseum / 1845–62

Barkas, T. P. (unknown) / private / Primitive Methodist Chapel, Earsdon; Lecture Room, Nelson Street, Newcastle / 1863

***Bartley, George** (ca. 1782–1858) / private; theater actor / English Opera House (Lyceum) / 1821–28

Bateman, Dr. (unknown) / private; possibly a medical doctor / a local hospital at Ilkley / 1852

Beechey, St. Vincent (1806–99) / Vicar of Worsley (1850–72) / Royal Manchester Institution / 1850–52

Berry, William (1774–1851) / private; schoolmaster and genealogist / Theatre Guernsey / 1811

***Bird, John** (d. 1840) / private; journeyman carpenter / local assembly venues including the town hall of Abingdon; public schools including Charter House, Westminster, and Eton; Russell Institution / 1814–40

Campbell, Captain (unknown) / Captain of Her Majesty's Indian Navy / Dovercourt Assembly Rooms, Harwich / 1866

Chalmers, Thomas (1780–1847) / Minister of Tron Church (1815–19) / Tron Church, Glasgow / 1815

***Children, Robert** (unknown) / private; bootmaker / local assembly venues including town halls of Oxford and Sudbury, Suffolk / 1835–44?

Dalton, John (1766–1844) / senior member of the Manchester Literary and Philosophical Society; FRS / Royal Institution / 1809

Davenport, Mr. (unknown) / unknown / Bowling Square Chapel, London (free lectures provided by the Society for Scientific, Useful, and Literary Information) / 1833

Faraday, Michael (1791–1867) / professor, Royal Institution (since 1833); FRS / Royal Institution / 1861

Freeman, Joseph (unknown) / private; schoolmaster / Baptist Chapel School, Great Ilford / 1856

Gladstone, John Hall (1827–1902) / FRS / Royal Institution / 1859

Goodacre, Mr. (unknown) / private / a riding school at Huddersfield / 1822

Grant, Robert (1814–92) / professor, University of Glasgow (since 1860); FRAS; FRS / Royal Institution / 1868–70

APPENDIX

Henderson, Ebenezer (1809–79) / private; clockmaker; FRAS / Liverpool and London, probably at London Mechanics' Institution / 1835–43?

Henry, Mr. (unknown) / unknown; self-described as a lecturer from the Polytechnic Institution / an unknown venue at Ipswich / 1846

Holden, Moses (1777–1864) / senior member of the Institution for the Diffusion of Knowledge in Preston and the BAAS; instrument maker / mechanics' institutes in northern English towns, especially in Preston / 1815–52

Howell, James (unknown) / private; clerk in the City of London / Adelphi Theatre; Her Majesty's Theatre / 1836–40?

Lardner, Dionysius (1793–1859) / editor and writer; member of the council of the BAAS (1838–40); FRS / place unknown, possibly institutions and theaters in London / ca. 1839

Lloyd, R. E. (unknown) / private; itinerant lecturer / Haymarket Theatre; Caledonian Theatre, Edinburgh; local assembly venues including the Music Room, Oxford / 1792–1829?

MacKeown, Mr. (unknown) / private / the Rev. M'Alister's meeting house at Holywood, Ireland / 1842

Moore, William P. (unknown) / headmaster of the Royal School, Cavan / Belturbet Literary and Scientific Society / between 1854 and 1855

Nichol, John Pringle (1804–59) / professor, University of Glasgow (since 1836) / Royal Manchester Institution; Sheffield Mechanics' Institute; Whittington Club and Metropolitan Athenaeum / 1849–51, 1858

Pepper, John Henry (1821–1900) / manager, Royal Polytechnic Institution (1858–72); FCS / Polytechnic Institution / 1862, 1865, 1870–71

Podmore, J. B. (unknown) / member of Jesus College, Cambridge / Birmingham Town Hall / 1857

Pond, John (1767–1836) / Astronomer Royal (1811–35); FRS; FRAS / Royal Institution / 1809–10

Popham, Charles (unknown) / private; itinerant lecturer / Philosophical Hall, Huddersfield; Theatre Royal, Birmingham; Theatre Royal, Leicester / 1850–51, 1853

Powell, Baden (1796–1860) / professor, University of Oxford (since 1827); FRS; FRAS / Royal Institution / 1850–51, 1858

Saull, William Devonshire (1783–1855) / private; radical philanthropist and geological collector / Chapel Court at Borough, London / 1832

Smyth, Charles Piazzi (1819–1900) / Astronomer Royal for Scotland (1846–88); FRS; FRAS; FRSE / Royal Institution / 1858

APPENDIX

Rogers, A. F. (unknown) / private; itinerant lecturer / local assembly venues in provincial towns including the Masonic Hall, Bath / 1824–25

Strutt, John William (1842–1919) / Senior Wrangler of Cambridge (1865); 3rd Baron Rayleigh (1873–1919); FRS / Witham Literary Institution / 1865

***Walker, Deane Franklin** (1778–1865) / private; experimental philosophy lecturer / English Opera House (Lyceum); King's Theatre; Strand Theatre; Colosseum; theaters and local assembly venues in provincial towns including Aberdeen Theatre, town hall of Oxford, and others / 1817–46?

***Walker, William** (bap. 1766, d. 1816) / private; experimental philosophy lecturer / Haymarket Theatre; theaters and local assembly venues in provincial towns including Lancaster Theatre, and others / 1782–1815

***Wallis, John** (1788–1852) / private; lecturer on contract basis / Royal Institution; London Mechanics' Institution; Russell Institution; Liverpool Mechanics' Institution; Royal Manchester Institution; Leeds Literary Institution, and others / 1825–51

Watson, Mr. (unknown) / unknown, probably private / Chelmsford Mechanics' Institution / 1837

Wheeler, J. (unknown) / private / Cosmorama Rooms, London / 1838

Young, Thomas (1773–1829) / professor of natural philosophy, Royal Institution (1801–3); FRS / Royal Institution / 1801

Notes

Introduction: A Sublime and Commercial Business

1. "Central Sun."

2. Chapman, *Victorian Amateur Astronomer*, 167. This article titled "London Lecturers: The Astronomical Mania" is a clipping from an unspecified newspaper dated sequentially to around 1839, in the Lee Album, M. H. S. Oxford Gunther 36, 1–3.

3. "Provincial Amusements," 442.

4. E. Henderson, *Treatise on Astronomy*, v, 9.

5. For example, Allan Chapman provides an overview of Victorian British "amateur" astronomers, in which he draws much on Herschel, Airy, and Ball; see Chapman, *Victorian Amateur Astronomer*. See also R. Jones, "Sir Robert Ball"; Lightman, *Victorian Popularizers of Science*, 397–421; and Hoskin, *Discoverers of the Universe*, ch. 11–13.

6. For examples of research on British cases of astronomical lectures in the eighteenth and nineteenth centuries, see Inkster, "Advocates and Audience"; Chapman, *Victorian Amateur Astronomer*, ch. 9; and Ruiz-Castell, "Astronomy and Its Audiences."

7. Cooter and Pumfrey, "Separate Spheres"; and O'Connor, "Reflections on Popular Science."

8. Lightman, *Victorian Popularizers of Science*, 12–13.

9. Gouyon, "1985, Scientists Can't Do Science Alone"; and Broks, *Understanding Popular Science*, 101–7.

10. J. Secord, *Victorian Sensation*, 437–38.

11. O'Connor, "Reflections on Popular Science," 336, 342–43, esp. fig. 2.

12. For examples of nineteenth-century British and transatlantic cases, see Pang, *Empire and the Sun*; Ratcliff, *Transit of Venus Enterprise*; and Nall, *News from Mars*.

13. Nall, *News from Mars*, 8.

14. For example, "Royal Polytechnic Institution," 193; and "Recollections of the Rev. John Eyton," 557.

15. Milner, *Gallery of Nature*, 1.

16. Ferguson, *Astronomy Explained*, 1.

17. *Oxford English Dictionary*, s.v. "sublime."

18. Whewell, *Astronomy*, 149.

19. Herschel, *Treatise on Astronomy*, 1–2.

20. *Oxford English Dictionary*, s.v. "sublime."

21. Burke, *Philosophical Enquiry*. For an introduction to the literature on the sublime in and beyond this period, see Kirwan, *Sublimity*; Shaw, *Sublime*; and Gross, *Scientific Sublime*, 1–22.

22. Golinski, "Sublime Astronomy," 137.

23. The "Bridgewater Treatises," sponsored by the bequest of the 8th Earl of Bridgewater, were a series of eight works exploring science on the basis of natural theology; see Topham, *Reading the Book of Nature*; and Fyfe, *Science and Salvation*.

24. Michael Faraday to Benjamin Abbott, June 1, 1813, in James, *Correspondence of Michael Faraday*, 56.

25. Garvey, *Silent Revolution*, 3; and J. Secord, *Victorian Sensation*, 522.

26. [Brougham], *Discourse of the Objects*, 6.

27. Royal Institution (hereafter RI): MM, vol. 1 (March 9, 1799), 1; and James, "Introduction," 1.

28. For discussion of the transformation of science in the nineteenth century, see J. Secord, *Visions of Science*; D. Knight, *Making of Modern Science*; and Russell, *Science and Social Change*. For a general introduction to astronomical disciplines, see Herrmann, *History of Astronomy*.

29. Ross, "Scientist."

30. J. Secord, *Victorian Sensation*, 24–38; Fyfe, *Steam-Powered Knowledge*; and Altick, *English Common Reader*.

31. Scott Bennett, "Revolutions in Thought"; Scott Bennett, "Editorial Character"; and Altick, *English Common Reader*, 332–39.

32. For *Vestiges*, see J. Secord, *Victorian Sensation*. For the Bridgewater Treatises, see Topham, "Science and Popular Education"; Topham, "Beyond the 'Common Context'"; Topham, *Reading the Book of Nature*; and Brooke

and Cantor, *Reconstructing Nature*, 156–58. For the RTS publications, see Fyfe, *Science and Salvation*.

33. See, for example, Gregory and Miller, *Science in Public*, 20–26.

34. Topham, "Publishing 'Popular Science,'" 135–37, esp. his analysis of the multiple meanings of the term "popular." See also Lightman, *Victorian Popularizers of Science*, 9–13.

35. Stewart, *Rise of Public Science*; Morton and Wess, *Public and Private Science*; and Elliott, "Birth of Public Science."

36. A Stranger in London, "A Visit to St. Paul's."

37. Altick, *Shows of London*, ch. 25 and 27.

38. Schaffer, "Natural Philosophy"; Golinski, "Noble Spectacle"; Morus, "Worlds of Wonder; and Wintroub, "Taking a Bow."

39. Fyfe and Lightman, "Science in the Marketplace."

40. J. Secord, *Victorian Sensation*, 437–39; and J. Secord, "Knowledge in Transit."

41. Lightman, *Victorian Popularizers of Science*, 9–13; and O'Connor, "Reflections on Popular Science."

42. Morus, "Replacing Victoria's Scientific Culture"; and Livingstone, *Putting Science in Its Place*.

43. Morus, "Replacing Victoria's Scientific Culture," 5.

44. Lightman, *Victorian Popularizers of Science*, viii.

45. Lightman, *Victorian Popularizers of Science*, 10.

46. Golinski, "Sublime Astronomy," 138.

47. Both Bartley and Bachhoffner have entries in the *Oxford Dictionary of National Biography*; see J. Knight, "Bartley, George"; and H. T. Wood, "Bachhoffner, George Henry." See also Morus, "'More the Aspect of Magic.'" For Bachhoffner's career at the Royal Polytechnic Institution, see Weeden, *Education of the Eye*.

Chapter 1. Pioneers

1. Inkster, "Advocates and Audience," 123.

2. Inkster, "Advocates and Audience," 119.

3. For Boyle's experimental philosophy and public demonstrations of experiments in the early history of the Royal Society, see Shapin and Schaffer, *Leviathan and the Air-Pump*; Golinski, "Noble Spectacle"; and Pumfrey, "Idea Above His Station."

4. For the cause and development of public lectures on Newtonian science in the early eighteenth century, see Stewart, *Rise of Public Science*; Mor-

ton and Wess, *Public and Private Science*; Elliott, "Birth of Public Science"; and Jacob and Stewart, *Practical Matter*, ch. 3.

5. "[Advertisement] For the Advancement of Natural Philosophy"; and Morton and Wess, *Public and Private Science*, 42.

6. Hodgson and Hauksbee, *Account of Hydrostatical and Pneumatical Experiments*.

7. Jacob and Stewart, *Practical Matter*, 80–83; and Fara, "Desaguliers, John Theophilus."

8. Desaguliers, *Physico-Mechanical Lectures*, 16; and King and Millburn, *Geared to the Stars*, 170–71.

9. Desaguliers, *Course of Experimental Philosophy*, 430–48, plates 31 and 32.

10. Wigelsworth, *Selling Science*, 177.

11. For example, to secure his reputation and income, Desaguliers had competed with two booksellers for publication of a textbook; see Wigelsworth, *Selling Science*, ch. 6.

12. Jacob and Stewart, *Practical Matter*, 68–69.

13. Elliott, "Birth of Public Science," 88.

14. Morton and Wess, *Public and Private Science*, 70, table I.

15. Morton and Wess, *Public and Private Science*, 77, table II.

16. Elliott, "Birth of Public Science," 88.

17. Morton and Wess, *Public and Private Science*, 93–109; and "George III: A Royal Passion for Science."

18. John Arden, *Short Account*, [1]. For a biographical study of John Arden, see Powers, *Philosopher Lecturing on the Orrery*.

19. John Arden, *Short Account*, [1–2].

20. James Arden, *Analysis of Mr. Arden's Course*, [2].

21. Bonnycastle, *Introduction to Astronomy*, 5.

22. "Proceedings of the Meeting," xiii. The quote is from Whewell's address to the British Association for the Advancement of Science.

23. Langford, *Polite and Commercial People*, 68–71. For the relations between polite manners and the construction of civilization, see also Thomas, *In Pursuit of Civility*.

24. Walters, "Conversation Pieces," 122–25; Morton and Wess, *Public and Private Science*, 61–64; and K. Taylor, "Mogg's Celestial Sphere."

25. Elliott, *Derby Philosophers*, 54–65. For the world's largest collection of Wright's works in Derby Museums, see "Joseph Wright of Derby."

26. Powers, *Philosopher Lecturing on the Orrery*.

27. B. Martin, *Young Gentleman*, 198.

28. B. Martin, *Description and Use of an Orrery*; and Millburn, "Benjamin Martin."

29. For further details of the development of orreries, see chapter 5 in this book.

30. Millburn and King, *Wheelwright of the Heavens*; Elliott, "Birth of Public Science," 89–91; and Rothman, "Ferguson, James."

31. Millburn, "Martin, Benjamin."

32. Lubbock, *Herschel Chronicle*, 59–60.

33. For further exploration of the everlasting ascendancy of some of the eighteenth-century authors, see chapter 4 in this book.

34. For biographical sources of Adam Walker, see Golinski, "Sublime Astronomy"; King and Millburn, *Geared to the Stars*, 309–11; and Carlyle, "Walker, Adam."

35. Elliott, "Birth of Public Science," 88; and Carlyle, "Walker, Adam."

36. Golinski, *Science as Public Culture*, 96–99.

37. P. M. Jones, *Industrial Enlightenment*, 74.

38. "Obituary of Eminent Persons"; and Carlyle, "Walker, Adam."

39. "Biographical Sketch of William Walker," 11. A copy of this magazine article is preserved in a scrapbook in the archive collections of the Royal Astronomical Society (hereafter RAS): Add MS 88: 122.

40. A. Walker, *Syllabus*, 2.

41. "Biographical Sketch of William Walker," 9.

42. King and Millburn, *Geared to the Stars*, 309–11; P. M. Jones, *Industrial Enlightenment*, 74; and "Biographical Sketch of William Walker," 9.

43. W. Walker, *Account of the Eidouranion*, 3–4.

44. Golinski, "Sublime Astronomy," 141–42.

45. W. Walker, *Account of the Eidouranion*, [1].

Chapter 2. Geography

1. Altick, *Shows of London*, 162.

2. Morus, *Frankenstein's Children*, 75–76. Ironically, the Adelaide Gallery's central location and the busy surrounding area was also a limitation that hindered its further development; see Altick, *Shows of London*, 377–82.

3. J. Secord, *Victorian Sensation*, 192–95.

4. On using the Natural History Museum at South Kensington as an example of a museum building's external iconography, see Livingstone, *Putting Science in Its Place*, 37–38. See also Forgan, "Building the Museum"; and Yanni, *Nature's Museums*.

5. Livingstone, *Putting Science in Its Place* provides a concise classic introduction of geographies of science. See also Powell, "Geographies of Science"; Finnegan, "Spatial Turn"; and Livingstone and Withers, *Geographies*. For background on notions of territory and social classes, see C. Smith and Agar, *Making Space for Science*.

6. Livingstone, *Putting Science in Its Place*, 184; and Morus, "Replacing Victoria's Scientific Culture," 9. The core idea of locality in the geographies of science is based on the influential work of Shapin and Schaffer in *Leviathan and the Air-Pump* (1985); see J. Secord, "Knowledge in Transit," 657.

7. Withers and Livingstone, "Thinking Geographically," 2.

8. J. Secord, "Knowledge in Transit."

9. Gieryn, "City as Truth-Spots," 5, esp. note 3. See also Gieryn, "Three Truth-Spots."

10. Lightman, "Refashioning the Spaces," 29–31, 42–43. For the prominent yet changing status of museums in nineteenth-century science, see Berkowitz and Lightman, *Science Museums*.

11. Finnegan, "Placing Science," 154–55. See also Livingstone, "Science, Site and Speech."

12. Dierig et al., "Introduction"; and Withers and Livingstone, "Thinking Geographically," 5–6.

13. Lach-Szyrma, *London Observed*, 17–18.

14. Lach-Szyrma, *London Observed*, 24.

15. For two examples of scholarly works that provide panoramic overviews of London's cultural and scientific landscapes in the nineteenth century, see Altick, *Shows of London*; and Fox, *London*. In "Scientific London," Morus et al. emphasize aspects of scientific life. For the development of the West End as an entertainment district in the nineteenth century, see McWilliam, *London's West End*. For examples of popular history depicting Victorian London, see Picard, *Victorian London*; and Flanders, *Victorian City*.

16. Bellon, "Science"; and Lightman, "Refashioning the Spaces," 25.

17. Wellbeloved, *London Lions*, [v].

18. J. Secord, *Victorian Sensation*, 178–79. For oral culture in nineteenth-century British society, see also J. Secord, "How Scientific Conversation Became Shop Talk.".

19. Altick, *Shows of London*, ch. 12; and R. D. Wood, "Diorama."

20. Wellbeloved, *London Lions*, 23–28.

21. "View of Roslyn Chapel," 132.

22. Lach-Szyrma, *London Observed*, 234.

23. Wellbeloved, *London Lions*, 49–53; and Altick, *Shows of London*, 246–48.

24. "Arcana of Science"; "Nota Bene"; and "Voice from Egyptian Hall."

25. Wellbeloved, *London Lions*, 79–81; Altick, *Shows of London*, 261–63; and Park and Park, "Goya's Living Skeleton."

26. Advertisement from the *Times*, January 13, 1839, quoted in L. Jackson, "Adelaide Gallery."

27. *Mogg's New Picture of London and Visitor's Guide to It [sic] Sights* (London, 1844), quoted in L. Jackson, "Adelaide Gallery." See also Altick, *Shows of London*, 377–82; "Drawing," inventory no. 1862,0614.689, British Museum, https://www.britishmuseum.org/collection/object/P_1862-0614-689.

28. Wellbeloved, *London Lions*, 1–2.

29. Wellbeloved, *London Lions*, 5–8. See also King and Millburn, *Geared to the Stars*, 339–40; and Busby, "Papers in Mechanics."

30. Timbs, *Curiosities of London*, 789.

31. Booth, *Theatre in the Victorian Age*, 5. Booth's statistics is based on an appendix from *Report from the Select Committee on Theatrical Licenses*, 295. For general situations of theaters in the Regency and Victorian eras, see also Mackintosh, "Departing Glories"; R. Jackson, *Victorian Theatre*; and Donohue, *Cambridge History of British Theatre*.

32. Booth, *Theatre in the Victorian Age*, 8, plate 1; and Mackintosh, "Pit, Boxes and Gallery."

33. Lach-Szyrma, *London Observed*, 213.

34. Grant, *Great Metropolis*, 90–91. See also R. Jackson, *Victorian Theatre*, 16–17; and Picard, *Victorian London*, 253–57.

35. Hays, "London Lecturing Empire"; and Morus et al., "Scientific London."

36. Poland was under Russian rule and had limited autonomy after the Congress of Vienna in 1815. Later in 1830 Lach-Szyrma played a central role in the November Uprising against Russia. He was exiled for life after the failure of the uprising and subsequently moved to England as a journalist and translator.

37. Lach-Szyrma, *London Observed*, 98.

38. Timbs, *Curiosities of London*, 532–33, 719.

39. Michael Faraday to Benjamin Abbott, June 1, 1813, in James, *Correspondence of Michael Faraday*, 56.

40. The Automatical Theatre was the automaton exhibition operated by Henri Maillardet in London between 1798 and 1817. See Altick, *Shows of London*, 350.

41. Michael Faraday to Benjamin Abbott, June 11, 1813, in James, *Correspondence of Michael Faraday*, 62.

42. *Report from the Select Committee on Dramatic Literature*, appendix no. 12, 249.

43. *Report from the Select Committee on Theatrical Licenses*, 295. John Timbs cites a different number for the Lyceum's capacity, giving it as 1,700; see Timbs, *Curiosities of London*, 789.

44. For the capacity of London theaters, see Timbs, *Curiosities of London*, 789; and *Report from the Select Committee on Theatrical Licenses*, 295. See also Mander and Mitchenson, *Theatres of London*. The Royal Institution began keeping attendance records in 1832, but theaters did not, so it is difficult to trace accurate numbers for audiences. Nevertheless, even if we assume a theater was half full, attendance at a theater lecture was still larger than at a scientific institution.

45. James, "Running the Royal Institution," 138–39, esp. table 6.2.

46. *Report from the Select Committee on Dramatic Literature*, 129.

47. "Biographical Sketch of William Walker," 10.

48. "Memoir of Stephen Kemble," 11. Benjamin Smart was among the most prominent authors and teachers of elocution in the early nineteenth century. Faraday had attended Smart's private lessons. See Morus, *Frankenstein's Children*, 20–21.

49. "Death of Mr. George Bartley."

50. An air is a songlike vocal or instrumental composition—a variant of an aria in opera. King and Millburn, *Geared to the Stars*, 312.

51. Wellbeloved, *London Lions*, 3.

52. See, for example, the advertisement for Lloyd's lecture in "Advertisements." See also Altick, *Shows of London*, 314–17.

53. "Adelphi"; "Advertisement"; and Chapman, *Victorian Amateur Astronomer*, 167.

54. It is debatable when Bartley started his Lenten astronomy lectures. The earliest advertisement I have found is dated April 1821. The producer Samuel James Arnold's letter to the Lord Chamberlain in January 1826, however, claims that the show was held in the English Opera House "afar the last seven years," which indicates that its debut was in 1819; British Library (hereafter BL): Add MS 42875, f. 444.

55. Kitchener, *Economy of the Eyes*, 166–67; and King and Millburn, *Geared to the Stars*, 317.

56. Hood, "Love and Lunacy," 79.

57. Chapman, *Victorian Amateur Astronomer*, 167.

58. Foulkes, *Church and Stage*, 32–34; and Altick, *Shows of London*, 364.

59. Foulkes, *Church and Stage*, 32. In *Report from the Select Committee on Dramatic Literature* (1832), John Payne Collier, the Lord Chamberlain's Examiner of Plays, claimed that many theaters continued playing during Lent; see also R. Jackson, *Victorian Theatre*, 18–19.

60. "King's Theatre."

61. *Metropolitan Magazine* 34 (February 1834): 56.

62. RAS: Add MS 88: 6.

63. "[Advertisement] Appropriate to Lent."

64. Hays, "London Lecturing Empire," 94. For the history of the Royal Institution, see James, *"Common Purposes of Life"*; Forgan, "Royal Institution of Great Britain"; and Berman, *Social Change*. For the London Institution, see Hays, "Science in the City."

65. Many scholarly works have discussed public science in English provincial cities or towns. See, for instance, P. M. Jones, *Industrial Enlightenment*; and Elliott, *Derby Philosophers*.

66. Morus et al., "Scientific London," 129.

67. Wilson and Geikie, *Memoir of Edward Forbes*, 324–25.

68. "Obituary: Daniel Moore"; and Hays, "London Lecturing Empire," 96. For Moore's involvement in the Royal Institution's committees and management, see RI: MS GB 1: 117, 147.

69. "New Scientific Establishment at Bath," 190. This proposed new institution was founded in 1824 as the Bath Literary and Scientific Institution and gained royal patronage in 1830. The institution was still in existence and operating as of 2024, when its bicentenary celebrations were observed. See "Inspiring Minds Since 1824."

70. Forgan, "'National Treasure House,'" 32–34.

71. RI: MM (March 9, 1799) 1: 1. See also James, "Introduction," 1.

72. Hays, "London Lecturing Empire," 94. Contemporary authors often mentioned the resources of these four institutions in correspondence and literature. See, for example, Charles Lamb's letter to Mr. Manning on February 26, 1808, in Talfourd, *Letters of Charles Lamb*, 186; and Timbs, *Curiosities of London*, 516–25. For the history of the London Institution, see Hays, "Science in the City"; and Kurzer, "Chemistry and Chemists." For the Surrey Institution, see Kurzer, "History of the Surrey Institution"; and Parolin, *Radical Spaces*, 190–98.

73. Parolin, *Radical Spaces*, ch. 6 and 7.

74. Hays, "Science in the City," 146n2.

75. Michael Faraday to Edward Magrath, July 23, 1826, in James, *Correspondence of Michael Faraday*, 417; and Morus, *Frankenstein's Children*, 30–31.

76. "Astronomy," *Journal of the Royal Institution*.

77. "Astronomy," *Journal of the Royal Institution*, 89, 108. In the nineteenth century, Ceres was classified first as a planet and later as the largest asteroid. In 2006 the International Astronomical Union reclassified Ceres as a dwarf planet.

78. Cantor, "Thomas Young's Lectures," 92–93; Berman, *Social Change*, 23–24; and James, "Introduction," 7.

79. After Baden Powell's death in 1860, his widow Henrietta Grace, to commemorate her late husband, and to set her surviving children apart from their half-siblings and cousins, styled the family name "Baden-Powell."

80. Corsi, "Powell, Baden."

81. Unpublished transcript of Babbage's course in the archives of the Royal Institution: Roberts, "Astronomy by Charles Babbage."

82. "Rotation of the Earth," May 17, 1851.

83. Previous scholarly works mentioning John Wallis include Hays, "London Lecturing Empire," 99; J. Secord, *Victorian Sensation*, 450–51; and James, *Christmas at the Royal Institution*, xvii.

84. "Proceedings of the Royal Institution"; RI: MS GB 2: 47/71C&D; and RI: MS LE 2: Index to Lectures 1829–41 and Attendance Figures, 127.

85. "Obituary." Wallis's correspondence with the Royal Manchester Institution in 1843 and 1845 shows his address as 338 Albany Road, Camberwell; see Manchester Archives (hereafter MA): M6/1/49/3/p162; and MA: M6/1/49/4/p13.

86. "City Public Lectures."

87. J. Secord, *Victorian Sensation*, 450.

88. "London Mechanics' Institution."

89. Wallis's lectures at the Royal Institution took place on December 27, and 29, 1838, and January 1, 3, 5, and 8, 1839; his lectures at the London Institution presented on December 17, 20, 24, 27, and 31, 1838, and January 3, 7, and 10, 1839. See "[Advertisement] London Institution"; and RI: MS LE 2: 127.

90. For Wallis's pay, see Hays, "London Lecturing Empire," 98–99. The London Mechanics' Institution paid Wallis 27 guineas for six lectures during the 1830s. He received higher pay (40 guineas for six lectures) from the London Institution.

91. MA: M6/1/49/2/p143; and MA: M6/1/49/3/p162. For other exam-

ples of extramural lecturers and their struggles for an institutional appointment, see Morrell, "Practical Chemistry"; and Anderson, "Chemistry Beyond the Academy."

92. RI: MS GB 2: 47/71C&D.

93. "Lectures and the Lecturers."

94. "Mechanics' Institution at Liverpool"; and "Lectures and the Lecturers." See also J. Secord, *Victorian Sensation*, 191–98.

95. Many scholarly works discuss the history of mechanics' institutes in Britain. See, for example, Royle, "Mechanics' Institutes"; M. D. Stephens and Roderick, "Science, the Working Classes and Mechanics' Institutes"; and Inkster, "Science and the Mechanics' Institutes." See also Wach, "Culture and the Middle Classes," 377–78n4, for a summary of the related bibliography.

96. Berman, *Social Change*, 100–101, 114–16. Berman nonetheless notes that the category of professionals does not matter sociologically to the formation of institutional governorship. He emphasizes that the "type," or ideological interest, of an individual, is more relevant to the analysis.

97. This is according to Forgan, "'National Treasure House,'" 32, esp. note 38. The price could vary depending on terms. For example, in 1850, a seasonal subscription to the lectures at the theater hall cost only 2 guineas, while the cost of subscriptions to both the theater hall and laboratory lectures was 3 guineas; see RI: GB 2: 99D. For more general description of members and audiences of the Royal Institution, see Forgan, "'National Treasure House,'" 31–40.

98. Russell, *Science and Social Change*, 151–52. For the construction of the Royal Institution and the changes to its architectural space throughout its history, see James and Peers, "Constructing Space."

99. MA: M6/1/70/64. The rules were valid as of March 1841.

100. "Lectures on Astronomy, by John Wallis."

101. "Mechanics' Institutions of This District."

102. W. C. Taylor, "Polytechnic Schools," 78–79.

103. "Chelmsford Mechanics' Institute."

104. Inkster, "Science and the Mechanics' Institutes."

105. Podmore, *Lecture on Astronomy*, title page, vi.

106. J. Secord, *Victorian Sensation*, 31, fig. 1.8, and 41–76. For the significance of steam-powered printing and publishing in the science marketplace, see Topham, "Publishing 'Popular Science'"; and Fyfe, *Steam-Powered Knowledge*.

107. For a consideration of the *Penny Magazine*, see Altick, *English Com-*

mon Reader, 270–71, 332–39; and Scott Bennett, "Editorial Character." For the *Magazine of Science*, see Sheets-Pyenson, "Popular Science Periodicals," 551, 567; and Huang, "From Grub Street to the Colony," 196–205.

108. "W. D. Saull." For the radical ideas and career of W. D. Saull, see Desmond, *Reign of the Beast*.

109. "Notice."

110. Podmore, *Lecture on Astronomy*, v–vi.

111. "Swimming."

112. J. Secord, *Victorian Sensation*, 46.

113. "Faith in Astronomy," 350.

114. Reid, *"Two Systems of Astronomy."*

115. The Science Museum Group (hereafter SMG): SCM-Astronomy: 1980-930/13.

116. "Astronomy," *Hampshire Advertiser*; "Sudbury"; and "Lectures on Astronomy," *Hampshire Advertiser*.

117. "Astronomical Lectures."

118. "March 1811: From the Newspapers," Priaulx Library, https://www.priaulxlibrary.co.uk/node/426. I thank Dinah Bott for this information. For a biography of William Berry, see Clark, "William Berry."

119. Moore, "Wonders of Astronomy."

120. "Lectures on Astronomy," *Belfast News-Letter*.

121. "Lecture on Astronomy."

122. "Ilford and Barking."

123. "Harwich: Lecture on Astronomy."

124. For example, the population of Sudbury was 7,969 (in 1841); of Ryde 11,795 (in 1841); of Ilkley 973 (in 1851); and of Great Ilford 4,523 (in 1851). The census data are from the website "A Vision of Britain Through Time," University of Portsmouth, http://www.visionofbritain.org.uk/. All the data cover either registration subdistrict or parish level unit.

125. "Central Sun," 99.

126. Chapman, *Victorian Amateur Astronomer*, 166.

127. "Newcastle upon Tyne."

128. The data on occupations is from the census of 1831. Population figures depend on what unit the census covers. The parish of Earsdon contained a few other neighboring townships so its population would be larger than the subdistrict of Earsdon. Earsdon is now a historical parish; after 1935, the original unit was split among several other civil parishes.

129. Lightman, "Refashioning the Spaces," 27, 31, 37–38. For the devel-

opment of Kew Gardens and everyday practices of botanists in the nineteenth century, see also Endersby, *Imperial Nature*.

130. Lightman, "Refashioning the Spaces," 26.

131. Finnegan, "Placing Science," 155.

CHAPTER 3. AFFILIATION

1. Martin Rudwick adopted the term "gentlemanly specialists" to denote the geologists of early nineteenth-century Britain; see Rudwick, *Great Devonian Controversy*. Similar terms such as "gentlemen of science" are often used in scholarship. See, for example, Morrell and Thackray, *Gentlemen of Science*; and Bowler and Morus, *Making Modern Science*, ch. 15, esp. 363–71.

2. John Pickstone uses the term "institutional sites of analysis" to designate the establishments that hosted scientific analysis functions. See Pickstone, *Ways of Knowing*, 130–34. See also Bowler and Morus, *Making Modern Science*, 363–71; Russell, *Science and Social Change*; and Cardwell, *Organisation of Science*.

3. Alberti, "Amateur and Professionals."

4. Ross, "Scientist," 65–88.

5. Porter, "Gentlemen and Geology," 810n5.

6. Chapman, *Victorian Amateur Astronomer*.

7. *Oxford English Dictionary*, s.v. "career (n.)."

8. Porter, "Gentlemen and Geology," 824, esp. note 70.

9. Hays, "London Lecturing Empire"; and Inkster, "Advocates and Audience."

10. Lightman, *Victorian Popularizers of Science*, viii, 489.

11. "Easter Amusements, &c."

12. "Eclipses," *Aberdeen Journal*, January 2, 1861; and "Astronomical Information."

13. RAS: Add MS 88: 4.

14. "Theatres, &c."

15. The baptism record at Edmonton shows Adams's birth date; he was baptized on April 24, 1803. The announcement of his death appeared in the *Pall Mall Gazette* on November 21, 1871, which indicates he was sixty-eight years old. His burial record is from the London Metropolitan Archives, All Saints, Edmonton, Register of Burials, DRO/040/A/01, item 023.

16. Cockburn et al., "Schools."

17. This is according to the marriage certificate and the census record in 1871.

18. "[Advertisement] Royal Adelphi Theatre, Strand."

19. "[Advertisement] Lyceum Theatre," *Morning Chronicle*, March 22, 1861. The year cited for Adams's lecture is sometimes inconsistent with the earliest playbill in 1830. For example, the advertisement in 1861 claimed this was the "31st year" of Adams's annual lecture, yet the debut should be 1831 if this claim was correct. It is not clear why Adams did not include his performance before 1831 on the list.

20. "Public Amusements."

21. "Science Gossip."

22. "Varieties: Mr. Adams' Lectures."

23. "Miscellaneous: Astronomical Lectures," 224.

24. "[Advertisement] Theatre Royal, Haymarket," *Theatrical Observer*.

25. "King's Theatre."

26. "Varieties: Mr. Adams's Orrery."

27. "Fine Arts."

28. The first edition of the Bridgewater Treatises was published in the 1830s, the same period during which this review appeared.

29. The rates are from the newspaper advertisement, "[Advertisement], Theatre Royal, Adelphi," *Morning Chronicle*.

30. The amount of the average wage is from Picard, *Victorian London*, appendix I. The price also equals the average weekly income of many other unskilled workers, such as coffee-stall keepers or female copy clerks. During the long Victorian era, the levels of wage and cost of living varied and could differ significantly between decades. The data cited here are appropriate for the mid-nineteenth century.

31. Mackintosh, "Pit, Boxes and Gallery," 553.

32. The National Standard Theatre (1837–1940), which Dickens called "The People's Theatre," was the largest establishment of its kind in London. Dickens wrote a series of articles on the leisure habits of the working classes in *Household Words* magazine; the one cited here was published on April 13, 1850. See R. Jackson, *Victorian Theatre*, 28.

33. See, for example, "[Advertisement] Mr. C. H. Adams's Orrery."

34. "Royal Polytechnic."

35. It was actually the thirty-second year if one starts counting from 1830. See "[Advertisement] Royal Adelphi Theatre, Strand."

36. "Amusements of Passion Week."

37. "Mr. Adams's Orrery."

38. Cockburn et al., "Schools."

39. "Destruction of Saville House."
40. "Royal Polytechnic."
41. "Literary and Art Gossip."
42. H. T. Wood, "Bachhoffner, George Henry."
43. Bachhoffner, *Chemistry*, ix.
44. It was founded as the Polytechnic Institution and the name was changed to the Royal Polytechnic Institution in 1841, when Prince Albert became the patron. For convenience, the two names are used interchangeably here.
45. Altick, *Shows of London*, 382. For further detail on the history of the Royal Polytechnic Institution, see Lightman, "Lecturing"; Morus, "'More the Aspect of Magic'"; and Weeden, *Education of the Eye*.
46. Weeden, *Education of the Eye*, 9–14; and Altick, *Shows of London*, 382.
47. Altick, *Shows of London*, 375–77. Although the Royal Polytechnic Institution was originally established for this educational purpose, the struggle of its path between scientific use and profitable entertainment had been a constant shadow over it. Such struggle was common among many contemporary spectacles. See, for example, J. Secord, on the Crystal Palace at Sydenham, "Monsters at the Crystal Palace."
48. Morus, "'More the Aspect of Magic,'" 352–54.
49. See, for example, "[Advertisement] Educational Classes."
50. See, for example, "[Advertisement] Royal Polytechnic Institution," *Examiner*; and "[Advertisement] Royal Polytechnic Institution," *Era*.
51. "Rotation of the Earth," May 3, 1851.
52. "[Advertisement] Royal Polytechnic Institution," *Literary Gazette*, no. 1914 (September 24, 1853): 921; and "[Advertisement] Royal Polytechnic Institution," *Literary Gazette*, no. 1938 (March 11, 1854): 217.
53. "[Advertisement] Re-Opening of the Royal Polytechnic Institution."
54. "Hydroelectric Machine"; and Morus, "'More the Aspect of Magic,'" 352.
55. "Hydro-electric Machine"; Weeden, *Education of the Eye*, 25; and Morus, "'More the Aspect of Magic,'" 352–53.
56. See "[Advertisement] Royal Polytechnic Institution," *Era*; and "[Advertisement] Theatre Royal, Adelphi," *Era*.
57. "Drama and Public Amusements."
58. For example, "[Advertisement] Theatre Royal, Adelphi" and "[Advertisement] Royal Polytechnic Institution," *Athenaeum*, March 31, 1849; "[Advertisement] Theatre Royal, Adelphi" and "[Advertisement] Royal Polytechnic Institution," *Athenaeum*, March 23, 1850; and "[Advertisement] Last

Night of the Orrery" and "[Advertisement] Royal Polytechnic Institution," *Daily News*, April 14, 1854.

59. H. T. Wood, "Bachhoffner, George Henry"; Altick, *Shows of London*, 161; and Weeden, *Education of the Eye*, 55. For the history of the Colosseum and its longtime financial struggle, see Altick, *Shows of London*, 141–62; and Lightman, "Science in Regent's Park."

60. "Royal Colosseum," December 25, 1856.

61. "Miscellaneous Exhibitions."

62. Altick, *Shows of London*, 161.

63. "[Advertisement] Royal Colosseum," *Era*.

64. For more discussion on John Henry Pepper, see Lightman, "Lecturing," 111–24; Cane, "John H. Pepper"; and J. Secord, "Quick and Magical Shaper."

65. "Colosseum." The charge of four shillings and sixpence was for day and evening admissions in total; see also Lightman, "Science in Regent's Park," 24.

66. "Royal Colosseum," December 6, 1857. This *Morning Chronicle* report simply mentions "the retirement of the other shareholders" and that Bachhoffner has "taken upon himself the entire management and burthen of the establishment." Another sign of the break of the company is from advertisements. The company's title, "Colosseum of Science and Art Company (Limited)," appeared in advertisements in early 1857, but it had disappeared by early 1858. Bachhoffner had been noted as the sole lessee since then. See, for example, "[Advertisement] Astronomy: Royal Colosseum"; and "[Advertisement] Royal Colosseum," *Standard*.

67. "London Exhibitions, &c."; and Lightman, "Science in Regent's Park," 32.

68. H. T. Wood, "Bachhoffner, George Henry."

69. "Atmosphere of the Metropolitan Railway"; "Arts and Manufactures"; and "Gas as Fuel."

70. *Mogg's New Picture of London and Visitor's Guide to It [sic] Sights* (London, 1844), in L. Jackson, "Adelaide Gallery." There is inconsistency in the title and publication year of this book between Jackson's and other online sources; for example, WorldCat shows the book is *Mogg's New Picture of London; or, Strangers' Guide to the British Metropolis* (London, 1845).

71. Chapman, *Victorian Amateur Astronomer*. Chapman compares this "Grand Amateurs" characteristic with the situations in Germany, France, and Russia.

72. Morrell, "Professionalisation," 980.

73. J. Secord, *Victorian Sensation*, 404.

74. J. Secord, *Victorian Sensation*, 403–10; Porter, "Gentlemen and Geology," 817–25; Bellon, "Joseph Dalton Hooker's Ideals"; and Endersby, *Imperial Nature*, 8–13, 249–75.

75. Chapman, *Victorian Amateur Astronomer*, 14–31. For the milieu of British astronomy during most of the nineteenth century, see also Hutchins, *British University Observatories*, ch. 2.

76. John Herschel's father, William, and aunt, Caroline, were both prominent astronomers. His mother was a rich widow when she married William. Many scholarly works have discussed the Herschel family in detail. See, for example, Holmes, *Age of Wonder*, ch. 2 and 4. For Lord Rosse's life and astronomical career, see articles collected in Morlan, *William Parsons*.

77. Chapman, "Science and the Public Good"; and Perkins, "'Extraneous Government Business.'"

78. Hutchins, *British University Observatories*, 59–61.

79. Hutchins, *British University Observatories*, 62–71, tables 2.1–2.3, 87. The term "Greenwich-Cambridge Axis" was coined by David W. Dewhirst to describe the elite network dominating nineteenth-century British astronomy. See also Chapman, *Victorian Amateur Astronomer*, 13–18; and Nall, *News from Mars*, 26–27.

80. F. K. Hunt, "Planet Watchers"; and Chapman, *Victorian Amateur Astronomer*, 146–57. For discussions covering an earlier period, see also Croarken, "Astronomical Labourers."

81. Hutchins, *British University Observatories*, 62–63, table 6.1.

82. The concept of the "invisible technician" was created by Steven Shapin and has been widely adopted or used as a quip by other scholars, such as Richard Sorrenson writing about the eighteenth-century instrument maker George Graham. See Shapin, "Invisible Technician"; and Sorrenson, "George Graham."

83. A copy of sunspot drawings preserved in the Royal Astronomical Society archives is likely attributable to Adams. The author was "Charles H. Adams" of Edmonton, who industriously recorded the change of sunspots between 1819 and 1823. It probably confirms Adams's interest in astronomy in his early life; RAS: Add MS 44.

84. Weeden, *Education of the Eye*, 24–25; Morus, "Currents from the Underworld"; and Morus, *Frankenstein's Children*, 99–124.

85. For examples of Herschel's and Airy's lectures, see Herschel, *Essays from Edinburgh*; and Airy, "On the Total Solar Eclipses."

86. Hays, "London Lecturing Empire," 91–92.

87. Inkster, "Advocates and Audience," 121.

88. An earlier classic discussing scientific performance sites refers to Altick, *Shows of London*. For the rich connections between science, performance, and sites in Victorian era, see Fyfe and Lightman, "Science in the Marketplace"; and Morus, "Worlds of Wonder."

89. Barton, "'Men of Science.'"

90. Desmond, "Redefining the X Axis," 4–5.

91. See, for example, Morus, "Replacing Victoria's Scientific Culture"; and Mussell, "Private Practice."

92. J. Secord, *Victorian Sensation*, 48–51.

93. Topham, "Science and Popular Education"; Brooke and Cantor, *Reconstructing Nature*, 153–61; and Fyfe, *Science and Salvation*.

94. J. Secord, *Victorian Sensation*, 437.

95. J. Secord, *Victorian Sensation*, 437–38.

96. Morrell, "Practical Chemistry," 69; Anderson, "Chemistry Beyond the Academy." See also Huang, "From Grub Street to the Colony." For a general introduction to the London lecturing scene, see Hays, "London Lecturing Empire"; and Morus et al., "Scientific London."

97. Endersby, *Imperial Nature*, 11–12.

98. Wilson and Geikie, *Memoir of Edward Forbes*, 324–25.

99. An unpublished transcript of this course with an introduction by C. J. D. Roberts is preserved in the Royal Institution archives. See Roberts, "Astronomy by Charles Babbage."

100. "Lecture by the Senior Wrangler."

101. This person is not to be confused with the instrument maker John Bird (1709–76), who was born in Bishop Auckland, County Durham. There is no evidence that the two Birds were related.

102. "Lecturer of the Old School."

103. "Country News," 78; and "Provincial Occurrences," 549.

104. Science Museum Group, "Handbill: Bird."

105. "Lecturer of the Old School," 677.

106. George Adams and his son, George Adams Jr., were both mathematical instrument makers to King George III. James Ferguson had received an annual pension of £50 from King George III since 1761. William Herschel was appointed as the king's astronomer after his discovery of Uranus.

107. Chapman, *Victorian Amateur Astronomer*, 172–73.

108. "Astronomy," *Hampshire Advertiser*.

109. "Sudbury."

110. Chapman, *Victorian Amateur Astronomer*, 173.

111. Drayton, *Nature's Government*, 85–94.

112. Smiles, *Self-Help*. For more examples of Victorian rhetoric of self-improvement, see J. Secord, *Victorian Sensation*, 336–63.

113. J. P. Smith, *On the Relation*, 327; and J. Secord, *Victorian Sensation*, 345.

Chapter 4. Subjects

1. Crop the Conjuror, "Hints to Lecturers," 509.

2. James, "Reporting Royal Institution Lectures, 1826–1867."

3. For the history of the Christmas Lectures, see James, *Christmas at the Royal Institution*; and "History of the Christmas Lectures."

4. RI: MS GB 2: 47/71C&D.

5. Baily, "On a Remarkable Phenomenon." See Littmann et al., *Totality*, 69–83.

6. C. H. Adams's lecture syllabus can be seen in many posters or advertisements in periodicals. For example, "[Advertisement] Theatre Royal, Haymarket," *Theatrical Observer*; and SMG: 1980-930/1.

7. See Adams's syllabi in his early years, for example, "[Advertisement] Royal Adelphi Theatre, Strand"; and SMG: 1980-930/2.

8. SMG: 1980-930/12; SMG: 1980-930/13; and RAS: Add MS 88: 8. See also King and Millburn, *Geared to the Stars*, 312, fig. 19.3.

9. SMG: 1980-930/12.

10. King and Millburn, *Geared to the Stars*, 312, fig. 19.3.

11. Bush, "Again with Feeling," 499, table 1.

12. C. F. Blunt, *Beauty of the Heavens*. I will further discuss Blunt and his illustrative works in chapter 5. For discussion of the relationship between astronomical lectures and other forms of media, see Bush and Huang, "New Stars on Screen."

13. *Report of the Third Meeting*, xiii.

14. Yeo, "Genius, Method, and Morality"; Gascoigne, "From Bentley to the Victorians"; and Higgitt, *Recreating Newton*.

15. Gascoigne, "From Bentley to the Victorians," 227–30; and Brooke, *Science and Religion*, 135–51.

16. "Faith in Astronomy," 350. This quotation is from the essay "Of Astronomy," in Godwin, *Thoughts on Man*.

17. "[Advertisement] Theatre Royal, Haymarket," *Theatrical Observer.*

18. Bush, "Again with Feeling," 499, table 1.

19. "Solar Eclipse," 23; "Late Solar Eclipse," *Morning Post*; and "Late Solar Eclipse," *Illustrated London News.*

20. For solar eclipse expeditions and nineteenth-century scientific culture, see Pang, *Empire and the Sun*; Aubin, "Eclipse Politics"; and Levitt, "'I Thought This Might Be of Interest.'"

21. "[Advertisement] Theatre Royal, Haymarket," *Theatrical Observer.*

22. "[Advertisement] Theatre Royal, Haymarket," *Standard.*

23. BL: Add MS 42875: f. 446.

24. RI: LE 4: 48. The average size of the Discourse audience is from James, "Running the Royal Institution," 139, table 6.2. See also James, "Reporting Royal Institution Lectures."

25. Airy, "On the Total Eclipse."

26. Smyth used an older spelling of the island; the modern spelling is "Tenerife."

27. Smyth, "Account of the Astronomical Experiment."

28. Smyth, "Account of the Astronomical Experiment," 497.

29. Faraday, "On Mr. Warren de la Rue's Photographic Eclipse Results."

30. RI: LE 4: 80, 99; and James, "Running the Royal Institution," table 6.2.

31. It is not clear which new planets were referred to in this advertisement. See "[Advertisement] Theatre Royal Haymarket," *Morning Chronicle.*

32. "Astronomical Amusement."

33. "[Advertisement] Great Solar Eclipse."

34. "Miscellaneous Intelligence: Suspected Existence."

35. "Miscellaneous Intelligence: A Supposed New Interior Planet."

36. For example, "Discoverer of the New Planet," *Daily News*; "Discoverer of the New Planet," *Bradford Observer*; "Discoverer of the New Planet," *Jackson's Oxford Journal*; and "Interesting Astronomical Episode."

37. Victoria and Albert Museum (hereafter V&A): S.1702-1995.

38. "Musical and Dramatic Gossip," 481.

39. Baum and Sheehan, *In Search of Planet Vulcan*; and Levenson, *Hunt for Vulcan.*

40. Ratcliff, *Transit of Venus Enterprise*, 49–54; and Nall, *News from Mars*, 13–17, 179–81.

41. Ratcliff, *Transit of Venus Enterprise*, 54.

42. Arnold, "Ouranologia"; and BL: Add MS 42875: ff. 443–93[b]. The quotations are from f. 447[b].

43. For more detail on Lord Chamberlain's censorship of British theaters and dramas, see Booth, *Theatre in the Victorian Age*, 145–49; and Foulkes, *Church and Stage*, ch. 2.

44. BL: Add MS 53702: ff. 106[b]–7.

45. BL: Add MS 42875: f. 444.

46. I would like to thank an anonymous referee for this suggestion.

47. RI: GB 1: 87, 108.

48. BL: Add MS 42875: ff. 448–50. For a full transcript of the *Ouranologia*, see Huang, *Ouranologia*.

49. BL: Add MS 42875: ff. 453[b]–54.

50. BL: Add MS 42875: ff. 469[b]–70[b].

51. BL: Add MS 42875: f. 478.

52. BL: Add MS 42875: f. 457.

53. BL: Add MS 42875: f. 469.

54. BL: Add MS 42875: ff. 448[b]–49.

55. Chapman, *Victorian Amateur Astronomer*, 168–70.

56. Young, *Night-Thoughts*, Night IX, lines 772–85, quoted from Gilfillan, *Young's Night Thoughts*.

57. Ferguson, *Astronomy Explained*, 1.

58. BL: Add MS 42875: f. 446.

59. [T. Dick,] *Solar System*, x.

60. Similar Christian discourse was declared regarding the exhibits of the Great Exhibition of 1851. See Cantor, *Religion and the Great Exhibition*, 128–43.

61. "Recollections of the Rev. John Eyton," 557.

62. Gascoigne, "From Bentley to the Victorians"; and Brooke, *Science and Religion*, 117–51. For an introduction to the clockwork universe metaphors in the Age of Enlightenment, see D. Knight, *Public Understanding of Science*, 13–28.

63. Hone, *Real or Constitutional House*, title page. See also J. Secord, *Visions of Science*, 6–7.

64. Fyfe, "Publishing and the Classics"; Topham, "Biology"; and Brooke and Cantor, *Reconstructing Nature*, 153–61.

65. Paley, *Natural Theology*, 378.

66. RAS: Add MS 88: 6.

67. RAS: Add MS 88: 8.

68. Brooke and Cantor, *Reconstructing Nature*, 184–95.

69. BL: Add MS 42875: f. 460, 470[b]; and D. F. Walker, *Epitome of Astronomy*, 32.

70. Whewell, *Astronomy*, 152–53. For the use of familiar objects connecting everyday life in nineteenth-century science education, see Keene, "Familiar Science."

71. BL: Add MS 42875: f. 447.

72. BL: Add MS 42875: f. 471[b].

73. RI: MS GB 2: 61/126C–126D.

74. MA: M6/1/70/106.

75. Bury, *Idea of Progress*, 5.

76. Schaffer, "Nebular Hypothesis"; and J. Secord, *Victorian Sensation*, 56–61.

77. Rudwick, *Earth's Deep History* provides an excellent introductory source on the historiography of the scientific discoveries of Earth's deep history.

78. J. Secord, *Victorian Sensation*, 57. See also Rudwick, *Earth's Deep History*; and O'Connor, *Earth on Show*.

79. Quoted from Schaffer, "Nebular Hypothesis," 134.

80. Ruse, "Evolution"; and Bowler, *Evolution*, esp. ch. 3 and 4, provides concise introductions on the history of evolution and the idea of social progress in a pre-Darwinian context.

81. Schaffer, "Nebular Hypothesis." See also Ogilvie, "Robert Chambers"; Numbers, *Creation by Natural Law*; and Brush, "Nebular Hypothesis."

82. Whewell, *Astronomy*, 191.

83. J. Secord, *Victorian Sensation*, 57–60.

84. Schaffer, "Nebular Hypothesis," 144–53.

85. Nichol, *Views of the Architecture*, 210.

86. MA: M6/1/70/104; M6/1/71/9; M6/1/71/71.

87. "[Advertisement] Professor Nichol, of Glasgow."

88. Numbers, *Creation by Natural Law*, 36; Schaffer, "Nebular Hypothesis," 145; and Finnegan, *Voice of Science*, 18–19.

89. Herschel, "Address." See also Schaffer, "Nebular Hypothesis"; and Hoskin, "John Herschel's Cosmology."

90. Herschel, "Address," xxxviii.

91. Wallis, *Brief Examination*; and J. Secord, *Victorian Sensation*, 406–7, 450–51.

92. Crowe, "Astronomy and Religion," 211. For the history of extraterrestrial life debates before the twentieth century, see also Crowe, *Extraterrestrial Life*; S. J. Dick, *Plurality of Worlds*; and Guthke, "Nightmare and Utopia."

93. Young, *Night-Thoughts*, Night IX, lines 778–79, quoted in Gilfillan, *Young's Night Thoughts*.

94. Ruse, "Introduction," 6.

95. Ruse, "Introduction," 6–7.

96. Ruse, "Introduction," 11–13.

97. Ruse, "Introduction," 20–24; and Brooke, "Natural Theology."

98. Crowe, *Extraterrestrial Life*, 351–52.

99. Walker, *Epitome of Astronomy*, 33. Although this is quoted from the last (31st) edition of Walker's syllabus in 1824, the text is almost the same in previous editions published since the late eighteenth century.

100. BL: Add MS 42875: f. 493; and Ferguson, *Astronomy Explained*, 6.

101. King and Millburn, *Geared to the Stars*, 338–39.

102. E. Henderson, *Treatise on Astronomy*, v.

103. E. Henderson, *Treatise on Astronomy*, 106–12.

104. E. Henderson, *Treatise on Astronomy*, 107.

105. E. Henderson, *Treatise on Astronomy*, 112.

Chapter 5. Apparatus

1. Sagan, "The Planets."

2. Golinski, "Sublime Astronomy," 137.

3. Walters, "Conversation Pieces," 145.

4. History of Science Museum, Oxford, inventory no. 97810.

5. King and Millburn, *Geared to the Stars*, xiii.

6. For the most comprehensive survey of the history of orreries and other astronomical clockwork devices to date, see King and Millburn, *Geared to the Stars*. John R. Millburn's works also provide careful investigations into the sources of the development of orreries since George Graham; see, for example, Millburn, "Benjamin Martin." For a more general introduction to orreries, see, for example, Bailey et al., " Human Orrery"; Buick, *Orreries, Clocks, and London Society*; and Vanhoutte, "Performing Astronomy."

7. Science Museum Group, "Orrery Made by John Rowley."

8. Desaguliers, *Course of Experimental Philosophy*, 431, and plate 31; and B. Martin, *Young Gentleman*, 199, and plate 19.

9. Taub, "'Grand' Orrery."

10. Deane, *Description of the Copernican System*, 90–91.

11. Adams, *Treatise*, [361], item no. 126.

12. Adams, *Treatise* [360]–[361], item no. 124 and 125.

13. Millburn, "Benjamin Martin," 393.

14. Picard, *Dr. Johnson's London*, 296.

15. B. Martin, *Description and Use of Both the Globes*, 178.

16. B. Martin, *Young Gentleman*, 25–34, and plate VI; B. Martin, *Description and Use of Both the Globes*, 178–86, and plate V; and Millburn, "Benjamin Martin," 385.

17. Science Museum Group, "Orrery Planetary Model by Benjamin Martin."

18. Millburn, "Benjamin Martin," 386–87, table 1.

19. Millburn, "Benjamin Martin," 388.

20. B. Martin, *Description and Use of an Orrery*, 28.

21. Millburn, "Benjamin Martin," 392. See also Harvard University et al., *Apparatus of Science*, 52–54; and Collection of Historical Scientific Instruments, Harvard University, "Grand Orrery."

22. Millburn, "Benjamin Martin," 398.

23. W. Jones, *Description and Use*, [iii].

24. W. Jones, *Description and Use*, 44; Millburn, "Benjamin Martin," 399.

25. Clifton, *Directory*, 154–55.

26. Royal Museums Greenwich, "Orrery," inventory no. AST1062.

27. [W. Jones], *Catalogue*, (1814), 8.

28. Royal Museums Greenwich, "Orrery," inventory no. AST1055.

29. [W. Jones], *Catalogue*, (1814), 8.

30. Royal Museums Greenwich, "Orrery," inventory no. AST1066; and Lacy, *Introduction to Astronomy*, 44.

31. Lacy, *Introduction to Astronomy*, 45.

32. For a general survey of the Great Exhibition, see Altick, *Shows of London*, esp. 456–60; Auerbach, *Great Exhibition*; Bellon, "Science"; and Cantor, *Religion and the Great Exhibition*.

33. Altick, *Shows of London*, 457; and 460, in which Altick quotes from [Ellis,] *Official Catalogue*, 112–13. See also Auerbach, *Great Exhibition*, 137.

34. "Town Talk and Table Talk."

35. [Untitled,] *Standard*.

36. [Untitled,] *Standard*. See also Warren, *The Lily and the Bee*, 140–55.

37. Strutt, *Tallis's History*, 243–48.

38. For the full list of the classes and juries, see *Reports by the Juries*.

39. Strutt, *Tallis's History*, 149. The surname of this working man in the *Official Catalogue* and the *Reports by the Juries* was registered as "Facy" rather than "Facey;" it is not clear whether it is misspelled in Tallis's compendium. According to the *Official Catalogue*, R. Facy lived at Wapping Wall in the East Ends of London (article no. 195). See [Ellis,] *Official Catalogue*, 66.

40. Strutt, *Tallis's History*, 148–49.

41. *Reports by the Juries*, 307.

42. Science Museum Group, "Orrery Planetary Model." The Newton & Company Limited was a partnership formed in 1851 between two distant cousins, William Edward Newton and Frederick Newton, to form an optical business. William was also the heir to the globe maker Newton & Son. See Stevenson, "W. E. and F. Newton."

43. *Reports by the Juries*, 307.

44. Strutt, *Tallis's History*, 245; and *Reports by the Juries*, 307.

45. King and Millburn, *Geared to the Stars*, 208–12.

46. King and Millburn, *Geared to the Stars*, 212.

47. For more detail on some outstanding nineteenth-century orreries, see King and Millburn, *Geared to the Stars*, 322–40.

48. For William Pearson's biography, see Gurman and Harratt, "Revd Dr William Pearson"; and Clerke, "Pearson, William."

49. Gurman and Harratt, "Revd Dr William Pearson," 279. For a modern reconstruction of this model, see Edinburgh Society of Model Engineers, "Replica of the Orrery Built by the Revd. William Pearson."

50. Gurman and Harratt, "Revd Dr William Pearson," 279–80; King and Millburn, *Geared to the Stars*, 334; and Science Museum Group, "Orrery Planetary Model Designed by William Pearson."

51. Clarke, *Reflections*, 139–40; and Grogans, "Fulton's Orrery."

52. King and Millburn, *Geared to the Stars*, 337–38.

53. RAS: Add MS 88: 7.

54. D. F. Walker, *Epitome of Astronomy*, 3.

55. D. F. Walker, *Epitome of Astronomy*, 4.

56. D. F. Walker, *Epitome of Astronomy*, 6.

57. E. Henderson, *Treatise on Astronomy*, 171.

58. E. Henderson, *Treatise on Astronomy*, 171–72.

59. Butterworth, "Astronomical Lantern Slides"; and Bird, "Enlightenment and Entertainment."

60. King and Millburn, *Geared to the Stars*, 310.

61. Golinski, "Sublime Astronomy," 146–47; Bush, "Astronomical Lantern Slide," 18–20; and Science Museum Group, "Mechanical Lantern Slide."

62. Huang, "From Grub Street to the Colony."

63. [Francis,] "Astronomical Illustrations," (1840).

64. [Francis], "Astronomical Illustrations," (1841).

65. "Varieties: Mr. Adams' Lectures."

66. Wellbeloved, *London Lions*, 1–2.

67. The Yale Center for British Art at Yale University, New Haven, Connecticut, and the History of Science Museum at Oxford University, England, each has a piece of this model in their collections. See Yale Center for British Art, "Elton's Miniature Transparent Orrery"; and History of Science Museum, University of Oxford, "Astronomical Demonstration Device." See also Golinski, "Sublime Astronomy," 155–56.

68. Golinski, "Sublime Astronomy," 155n63; and Yale Center for British Art, "Elton's Miniature Transparent Orrery."

69. Golinski, "Sublime Astronomy," 155.

70. Pearson, "Planetary Machines," in *Cyclopaedia*.

71. Pearson, "Planetary Machines," in *Edinburgh Encyclopaedia*, 626.

72. Herschel. *Treatise on Astronomy*, 287.

73. Michael Faraday to Benjamin Abbott, June 1, 1813, in James, *Correspondence of Michael Faraday*, 56.

74. R. D. Wood, "Diorama in Great Britain"; and Altick, *Shows of London*, 163–72.

75. Lach-Szyrma, *London Observed*, 234.

76. Altick, *Shows of London*, 174, 198–210.

77. For a general survey of the history of the magic lantern, see Crangle et al., *Realms of Light*; Shepard, "Magic Lantern Slide"; and Magic Lantern Society, "Brief History."

78. A Mere Phantom, *Magic Lantern*. See also Heard, *Phantasmagoria*.

79. Butterworth, "Astronomy," 19.

80. Wellbeloved, *London Lions*, 2.

81. [W. Jones], *Catalogue*, (1797), 3–4.

82. Butterworth, "Astronomy," 19–20; and Wells, "Fleas."

83. E. Henderson, *Treatise on Astronomy*, 9.

84. See, for example, "[Advertisement] Royal Polytechnic Institution," April 8, 1854; and V&A: S. 1702-1995.

85. Chapman, *Victorian Amateur Astronomer*, 178. For examples of newspaper reports, see "Chelmsford Mechanics' Institute"; and "Phrenological Society." Both newspaper reports mentioned the use of the phantasmagoria at local lectures on astronomy.

86. Butterworth, "Lantern Tour."

87. Bush, "Again with Feeling."

88. Royal Museums Greenwich, "Lantern Slide."

89. This range of price values is according to a survey of catalogs and ad-

vertisements between 1815 and 1845. I would like to thank Martin Bush for providing this information.

90. The price is from "[Advertisement] The Beauty of the Heavens"; it was almost sixfold more expensive than the price of the first five editions of *Vestiges of the Natural History of Creation* (7s. 6d.), which was also a commercial success in the mid-1840s. The price of *The Beauty of the Heavens* was nonetheless reduced to about 1 guinea in later editions or reprints.

91. For a brief biography about the early career of Charles F. Blunt, see Bush and Huang, "New Stars on Screen."

92. B. E. Blunt, *Uranographia*, i.

93. C. F. Blunt, *Beauty of the Heavens*, v.

94. C. F. Blunt, *Beauty of the Heavens*, v.

95. C. F. Blunt, *Beauty of the Heavens*, 14, scene 2.

96. On the seasons, C. F. Blunt, *Beauty of the Heavens*, 81–82, scenes 63 and 64; and on the tides, 83, scene 66.

97. C. F. Blunt, *Beauty of the Heavens*, 78–79, scene 58.

98. A. Smith, *Smith's Illustrated Astronomy*, 6.

99. A Mere Phantom, *Magic Lantern*, 18–19. See also V. Hunt, "Raising a Modern Ghost."

100. Butterworth, "Lantern Tour."

101. Millburn, "Benjamin Martin," 398n88.

Chapter 6. Audiences

1. Fyfe, *Science and Salvation*, 144–45.

2. Susan Bennett, *Theatre Audiences*; and Fortier, *Theory/Theatre*, 87–100.

3. Gregory and Miller, *Science in Public*, 89–90; Locke, "Golem Science"; Broks, *Understanding Popular Science*, 122–23.

4. J. Secord, *Victorian Sensation*, 518.

5. Topham, "Beyond the 'Common Context,'" 261. For further discussion of the Bridgewater Treatises, see Topham, *Reading the Book of Nature*.

6. On Faraday's day-to-day preparatory work and lecturing, see James, "Running the Royal Institution"; and Morus, *Frankenstein's Children*. For analysis of members and audiences of the Royal Institution in the nineteenth century, see Berman, *Social Change*; Forgan, "'National Treasure House'"; and Lloyd, "Rulers of Opinion."

7. Lightman, *Victorian Popularizers of Science*, 227; and Page, *Earth's Crust*, iii.

8. J. Secord, *Victorian Sensation*, 136–37; and Fyfe, *Science and Salvation*, 157–58.

9. Altick, *English Common Reader*, 166–72; and Fyfe, *Science and Salvation*, 26–27.

10. Vincent, *Literacy and Popular Culture*, 27, fig. 2.2.

11. Vincent, *Literacy and Popular Culture*, 22; and Lightman, *Victorian Popularizers of Science*, 18.

12. For the Royal Institution, see Berman, *Social Change*; and Forgan, "National Treasure House." For the Great Exhibition, see Cantor, *Great Exhibition*, 259–68. For Victorian collections of natural history, see Alberti, "Museum Affect."

13. Lloyd, "Rulers of Opinion."

14. M. D. Stephens and Roderick, "Science"; and Lightman, "Science in Regent's Park," 25–26, 33–35.

15. Berman, *Social Change*, 100–101, 114–16.

16. Booth, *Theatre in the Victorian Age*, 1–26; and R. Jackson, *Victorian Theatre*, 9–17.

17. Picard, *Victorian London*, appendix I.

18. "Lecturer of the Old School," 677.

19. Morton and Wess, *Public and Private Science*; Walters, "Tools of Enlightenment"; and Walters, "Conversation Pieces." See also J. Secord, "Newton in the Nursery" for the case of the eighteenth-century author Tom Telescope and his popular book tailored for young gentlemen and ladies on Newtonian science.

20. W. C. Taylor, "Polytechnic Schools."

21. For the general situation of education in nineteenth-century Britain, see Kelly, *History of Adult Education*; and W. B. Stephens, *Education in Britain*.

22. K. Taylor, "Mogg's Celestial Sphere"; and Keene, "Playing Among the Stars."

23. Wellbeloved, *London Lions*, 4.

24. "Punch's Almanack for 1855"; and Rudwick, *Scenes from Deep Time*, 145, fig. 64.

25. "Punch's Almanack for 1866."

26. Lamb, "The Old and the New," 495. See also Altick, *Shows of London*, 228.

27. C. F. Blunt, *Beauty of the Heavens*, vi.

28. Keene, "Playing Among the Stars."

29. Lightman, *Victorian Popularizers of Science*, 15.

30. For examples of pioneering studies and writing about female naturalists and authors of science, see Shteir, *Cultivating Women*; and Gates, *Kindred Nature*. For biographical studies of women in astronomy, see Ogilvie, "Obligatory Amateurs"; Neeley, *Mary Somerville*; Brück, *Agnes Mary Clerke*; and Brück, *Women*.

31. Higgitt and Withers, "Science and Sociability"; and Lloyd, "Rulers of Opinion." See also Morrell and Thackray, *Gentlemen of Science*, 148–57.

32. Lloyd, "Rulers of Opinion," 11–12; and Golinski, *Science as Public Culture*, 194–95.

33. Higgitt and Withers, "Science and Sociability," 25.

34. RAS: Add MS 88: 35. Little is known about Mr. Perini and his planetarium. The Italian astronomer N. Perini built this new amusement in London around the late 1870s, and he was elected Fellow of the Royal Astronomical Society in 1880. See "Meeting of the Royal Astronomical Society."

35. Chapman, *Victorian Amateur Astronomer*, 273–93; and Lightman, *Victorian Popularizers of Science*, 21–23, 97–99, 469–88.

36. Hollingshead, "At the Play."

37. Hollingshead, "At the Play," 89.

38. Hollingshead, "At the Play," 89.

39. Hollingshead, "At the Play," 89–92.

40. "Lecture in Lent."

41. For studies of the image of scientists in popular culture and media, see Haynes, *From Faust to Strangelove*; and Frayling, *Mad, Bad, and Dangerous?*

42. Dickens, "Uncommercial Traveller," 349. See also Altick, *Shows of London*, 365.

43. Although *Father and Son* is well-received as literature of Victorian adolescence, modern biographers reject the stern portrait of Philip Henry Gosse in this memoir and note Edmund Gosse's partial depictions of his father. See, for example, Thwaite, *Glimpses of the Wonderful*; Lee, "Writing Victorian Lives"; and H. Henderson, *Victorian Self*, 117–58.

44. Wyld, *Notes*, 6–18. However, it is not clear whether Wyld arranged astronomical materials in the actual exhibition. See also Altick, *Shows of London*, 464–67; King and Millburn, *Geared to the Stars*, 320–21; and Lightman, "Spectacle in Leicester Square."

45. Gosse, *Father and Son*, 60.

46. Topham, "Beyond the 'Common Context,'" 249.

47. Topham, "Beyond the 'Common Context,'" 235.

48. J. Secord, *Victorian Sensation*, 338–42.

49. "Punch's Information for the People."

50. J. Secord, *Victorian Sensation*, 157–66.

51. Alberti, "Conversaziones"; and J. Secord, "How Scientific Conversation Became Shop Talk." For the significance of conversation about science and astronomy in polite culture in eighteenth-century Britain, see Walters, "Conversation Pieces."

52. J. Secord, *Victorian Sensation*, 178–79.

53. Poole, "Discourse of Bores."

54. Poole, "Discourse of Bores," 396.

55. Poole, "Discourse of Bores," 561.

56. Steele, *Englishman*, 72–73. See also King and Millburn, *Geared to the Stars*, 154; and Walters, "Conversation Pieces," 121–22.

57. Altick, *Shows of London*, 227.

58. A. Secord, "Botany on a Plate," 30. Many other studies also emphasize the concept of "pleasure" or "amusement" in eighteenth- and early nineteenth-century education. See, for example, Kohlstedt, "Parlors, Primers, and Public Schooling"; Riskin, "Amusing Physics"; and Keene, "Playing Among the Stars."

59. "Deformito-Mania." See also Altick, *Shows of London*, 253–67.

60. Chapman, *Victorian Amateur Astronomer*, 167. The article titled "London Lecturers: The Astronomical Mania" is a clipping from an unspecified newspaper dated sequentially to around 1839, in the Lee Album, M. H. S. Oxford Gunther 36, 1–3.

61. Chapman, *Victorian Amateur Astronomer*, 167.

62. Hood, "Love and Lunacy."

63. Jerrold, *Thomas Hood*, 288–90.

64. Hood, "Love and Lunacy," 37–38.

65. Hood, "Love and Lunacy," 79.

66. Hollingshead, "At the Play," 87.

67. "Provincial Amusements," 441.

68. For a case study of the relations between politics and theories of evolution in nineteenth-century London, see Desmond, *Politics of Evolution*.

69. The identity and gender of this "Lover of Truth" is, of course, unknown. For the convenience of writing, I use only male pronouns here.

70. Lover of Truth, "Copy of a Letter," 88.

71. Lover of Truth, "Copy of a Letter," 89.

72. Lover of Truth, "Copy of a Letter," 89.

73. King and Millburn, *Geared to the Stars*, 317. The full name of A. F. Rogers is unknown.

74. P. W. Martin, "Carlile, Richard." For a detailed biography of Carlile, see Aldred, *Richard Carlile*.

75. Carlile, *Address to Men of Science*.

76. Carlile, "Astronomers and Astrologers," 92.

77. Carlile, *Address to Men of Science*, 3.

78. Carlile, *Address to Men of Science*, 29.

79. Carlile, *Address to Men of Science*, 29.

80. Reid, "*Two Systems of Astronomy*," 162–63.

81. Reid, "*Two Systems of Astronomy*," 166–68.

82. J. W. S., "Wonders of the Heavens," 267.

83. Schaffer, "Nebular Hypothesis," 156.

Epilogue: An Enduring Legacy

1. "[Advertisement] To Lecturers," July 15, 1865.

2. "[Advertisement] To Lecturers," November 25, 1865.

3. "Mr. Deane Walker."

4. See, for example, "[Advertisement] Polytechnic"; and "[Advertisement] Royal Polytechnic."

5. "Royal Polytechnic."

6. Lightman, "Pepper, John Henry."

7. Altick, *Shows of London*, 470–73, 504–9.

8. On the significant influence of railways on Victorian culture and everyday life, see Freeman, *Railways*. See also J. Secord, *Victorian Sensation*, 24–28; and Fyfe, *Steam-Powered Knowledge*, 97–109.

9. For an analysis of the Crystal Palace at Sydenham and its tangle of spectacle, commerce and expertise, see J. Secord, "Monsters at the Crystal Palace." On the rise of public museums and museum visitors' experience in the second half of the nineteenth century, see Forgan, "Architecture of Display"; Forgan, "Building the Museum"; and Alberti, "Museum Affect."

10. Finnegan, *Voice of Science*, 94–129; and Nall, *News from Mars*, 25–31.

11. Mussell, "Arthur Cowper Ranyard"; and Lightman, *Victorian Popularizers of Science*, 295–351.

12. "[Advertisement] Royal Institution of Great Britain," April 20, 1868; "[Advertisement] Royal Institution of Great Britain," March 31, 1869; and "[Advertisement] Royal Institution of Great Britain," April 16, 1870. For Grant's biography, see Clerke, "Grant, Robert (1814–92)."

13. Lightman, *Victorian Popularizers of Science*, 397–417; and Butterworth, "Lantern Tour."

14. Inkster, "Advocates and Audience," 123.

15. Lightman, *Victorian Popularizers of Science*, 494–96.

16. Barton, "'Men of Science'"; and Chapman, *Victorian Amateur Astronomer*.

17. Dewhirst, "Greenwich-Cambridge Axis."

18. Lightman, *Victorian Popularizers of Science*, 5–8.

19. Golinski, "Sublime Astronomy," 138, 149–50.

20. Gross, *Scientific Sublime*.

21. Gross, *Scientific Sublime*, 263.

22. On the spread and prevalence of scientific naturalism in the late nineteenth century, see Lightman and Reidy, *Age of Scientific Naturalism*.

23. On the development of planetarium projectors and their position in the family tree of astronomical visual aids, see King and Millburn, *Geared to the Stars*, 341–68; Marché, *Theaters of Time and Space*; and Buick, *Orreries, Clocks, and London Society*, 311–33.

24. On orreries, performance, and the theater across media, see also Vanhoutte and Bigg, "On the Border"; and Vanhoutte, "Performing Astronomy."

25. Sagan, *Cosmos*, xvi.

BIBLIOGRAPHY

MANUSCRIPT AND DATABASE COLLECTIONS

British Library (BL), Manuscript Collections,.
British Library Newspapers (BLN), Gale Primary Sources (GPS). https://www.gale.com/primary-sources/british-library-newspapers.
British Periodicals (BP), ProQuest (PQ). https://about.proquest.com/en/products-services/british_periodicals/.
Eighteenth Century Collections Online (18th Century), GPS. https://www.gale.com/primary-sources/eighteenth-century-collections-online.
Illustrated London News Historical Archives, 1842-2003 (ILNHA), GPS. https://www.gale.com/c/illustrated-london-news-historical-archive.
London Metropolitan Archives, All Saints, Edmonton, Register of Burials.
Manchester Libraries, Information and Archives (MA).
Punch Historical Archive, 1841-1992, GPS. https://www.gale.com/c/punch-historical-archive.
Royal Astronomical Society Archives (RAS).
Royal Institution of Great Britain Archives (RI).
Royal Museums Greenwich, Astronomical and Navigational Instruments Collections. https://www.rmg.co.uk/collections.
Science in the Nineteenth-Century Periodical Index (SciPer). https://www.sciper.org.
Science Museum Group (SMG), Astronomy Collections; Art Collections. https://collection.sciencemuseumgroup.org.uk.
Victoria and Albert Museum (V&A), H. Beard Print Collections; Theatre and Performance Collections. https://www.vam.ac.uk/collections.

BIBLIOGRAPHY

PRIMARY AND SECONDARY SOURCES

Adams, George. *A Treatise Describing the Construction, and Explaining the Use, of New Celestial and Terrestrial Globes*, 3rd ed. London, 1772.

"Adelphi." *Theatrical Observer*, March 30, 1836, [1] (BP PQ).

"[Advertisement] Appropriate to Lent." *Leicester Chronicle*, February 26, 1853 (BLN GPS).

"[Advertisement] Astronomy: Royal Colosseum." *Morning Chronicle*, March 11, 1857 (BLN GPS).

"[Advertisement] The Beauty of the Heavens." *Examiner*, January 12, 1840 (BLN GPS).

"[Advertisement] The Educational Classes of the Royal Polytechnic Institution." *Examiner*, no. 1808 (September 24, 1842): 624 (BP PQ).

"[Advertisement] For the Advancement of Natural Philosophy and Astronomy." *Daily Courant*, January 11, 1705 (18th Century GPS).

"[Advertisement] Great Solar Eclipse." *Era*, July 8, 1860 (BLN GPS).

"[Advertisement] Last Night of the Orrery." *Daily News*, April 14, 1854.

"[Advertisement] London Institution." *Morning Chronicle*, November 5, 1838 (BLN GPS).

"[Advertisement] Lyceum Theatre." *Morning Chronicle*, March 22, 1861 (BLN GPS).

"[Advertisement] Mr. C. H. Adams's Orrery." *Essex Standard*, August 12, 1853 (BLN GPS).

"[Advertisement] Polytechnic." *Era*, April 13, 1862 (BLN GPS).

"[Advertisement] Professor Nichol, of Glasgow." *Athenaeum*, no. 1117 (March 24, 1849): 289 (BP PQ).

"[Advertisement] Re-Opening of the Royal Polytechnic Institution." *Era*, December 7, 1851 (BLN GPS).

"[Advertisement] Royal Adelphi Theatre, Strand" *Theatrical Observer*, April 6, 1830 (BP PQ).

"[Advertisement] Royal Colosseum." *Era*, February 22, 1857 (BLN GPS).

"[Advertisement] Royal Colosseum." *Standard*, February 19, 1858 (BLN GPS).

"[Advertisement] Royal Institution of Great Britain." *Pall Mall Gazette*, April 20, 1868 (BLN GPS).

"[Advertisement] Royal Institution of Great Britain." *Pall Mall Gazette*, March 31, 1869 (BLN GPS).

"[Advertisement] Royal Institution of Great Britain." *Pall Mall Gazette*, April 16, 1870 (BLN GPS).

"[Advertisement] Royal Polytechnic." *Standard*, March 29, 1865 (BLN GPS).

"[Advertisement] Royal Polytechnic Institution." *Athenaeum*, no. 1118 (March 31, 1849), 332.

"[Advertisement] Royal Polytechnic Institution." *Athenaeum*, no. 1169 (March 23, 1850), 314.

"[Advertisement] Royal Polytechnic Institution." *Athenaeum*, no. 1380 (April 8, 1854): 441.

"[Advertisement] Royal Polytechnic Institution." *Daily News*, April 14, 1854.

"[Advertisement] Royal Polytechnic Institution." *Era*, March 16, 1845 (BLN GPS).

"[Advertisement] Royal Polytechnic Institution." *Examiner*, no. 1933 (February 15, 1845): 109 (BLN GPS).

"[Advertisement] Royal Polytechnic Institution." *Literary Gazette*, no. 1914 (September 24, 1853): 921 (BP PQ).

"[Advertisement] Royal Polytechnic Institution." *Literary Gazette*, no. 1938 (March 11, 1854): 217 (BP PQ).

"[Advertisement] Theatre Royal, Adelphi." *Athenaeum*, no. 1118 (March 31, 1849), 332.

"[Advertisement] Theatre Royal, Adelphi." *Athenaeum*, no. 1169 (March 23, 1850), 314.

"[Advertisement] Theatre Royal, Adelphi." *Era*, March 16, 1845 (BLN GPS).

"[Advertisement] Theatre Royal, Adelphi." *Morning Chronicle*, March 21, 1853 (BLN GPS).

"[Advertisement] Theatre Royal Haymarket." *Morning Chronicle*, April 14, 1851 (BLN GPS).

"[Advertisement] Theatre Royal, Haymarket." *Standard*, March 30, 1858 (BLN GPS).

"[Advertisement] Theatre Royal, Haymarket." *Theatrical Observer*, February 27, 1839 (BP PQ).

"[Advertisement] To Lecturers, Institutions, and Persons Seeking a Profession." *Athenaeum*, no. 1968 (July 15, 1865): 65.

"[Advertisement] To Lecturers, Institutions, and Persons Seeking a Profession, at Home or Abroad." *Athenaeum*, no. 1987 (November 25, 1865): 709.

"Advertisement." *Theatrical Observer*, February 27, 1839, [3] (BP PQ).

"Advertisements." *Edinburgh Literary Journal*, May 30, 1829, 20.

BIBLIOGRAPHY

Airy, G. B. "On the Total Solar Eclipse of 1851, July 28." *Notices of the Proceedings at the Meetings of Members of the Royal Institution of Great Britain* 1 (May 2, 1851 [1851–54]): 62–68.

Alberti, Samuel J. M. M. "Amateur and Professionals in One County: Biology and Natural History in Late Victorian Yorkshire." *Journal of the History of Biology* 34, no. 1 (Spring 2001): 115–47.

Alberti, Samuel J. M. M. "Conversaziones and the Experience of Science in Victorian England." *Journal of Victorian Culture* 8, no. 2 (January 2003): 208–30.

Alberti, Samuel J. M. M. "The Museum Affect: Visiting Collections of Anatomy and Natural History." In Fyfe and Lightman, *Science in the Marketplace*, 371–403.

Aldred, Guy A. *Richard Carlile, Agitator: His Life and Times*. Pioneer Press, 1923.

Alexander, Denis, and Ronald Numbers, eds. *Biology and Ideology from Descartes to Dawkins*. University of Chicago Press, 2010.

Altick, Richard D. *The English Common Reader: A Social History of the Mass Reading Public, 1800–1900*, 2nd ed. Ohio State University Press, 1998.

Altick, Richard D. *The Shows of London*. Belknap Press of Harvard University Press, 1978.

"Amusements of Passion Week." *Era*, March 31, 1861 (BLN GPS).

Anderson, Robert. "Chemistry Beyond the Academy: Diversity in Scotland in the Early Nineteenth Century." *Ambix* 57, no. 1 (2010): 84–103.

"Arcana of Science: Museum of Natural History." *Mirror of Literature, Amusement, and Instruction* 10 (1828): 432.

Arden, James. *Analysis of Mr. Arden's Course of Lectures on Natural and Experimental Philosophy*. Printed for, and sold by the author, 1774.

Arden, John. *A Short Account of a Course of Natural and Experimental Philosophy*. Coventry: Printed by J. W. Piercy, in Broad-Gate, 1772.

Arnold, Samuel James. "Ouranologia." Manuscript submitted to the Lord Chamberlain's Office. January, 1826.

"Arts and Manufactures: New Patents." *Birmingham Daily Post*, September 30, 1871.

"Astronomical Amusement." *Punch* 14 (1848): 183.

"Astronomical Information, from Punch's Almanack for 2417: Eclipses." *Punch* 40 (1861): "Punch's Almanack for 1861."

"Astronomical Lectures: Riding-School, Huddersfield." *Leeds Mercury*, August 31, 1822 (BLN GPS).

BIBLIOGRAPHY

"Astronomy." *Hampshire Advertiser*, July 23, 1836 (BLN GPS).

"Astronomy." *Journal of the Royal Institution of Great Britain* 1 (1802): 86–89; 108–9.

"The Atmosphere of the Metropolitan Railway." *Daily News*, October 31, 1867.

Aubin, David. "Eclipse Politics in France and Thailand, 1868." In Aubin et al., *Heavens on Earth*, 86–117.

Aubin, David, Charlotte Bigg, and H. Otto Sibum, eds. *The Heavens on Earth: Observatories and Astronomy in Nineteenth-Century Science and Culture*. Duke University Press, 2010.

Auerbach, Jeffery A. *The Great Exhibition of 1851: A Nation on Display*. Yale University Press, 1999.

Bachhoffner, George H. *Chemistry as Applied to the Fine Arts*. London: J. Carpenter and Company, 1837.

Bailey, M. E., D. J. Asher, and A. A. Christou. "The Human Orrery: Ground-Based Astronomy for All." *Astronomy & Geophysics* 46, no. 3 (June 2005): 3.31–3.45.

Baily, Francis. "On a Remarkable Phenomenon That Occurs in Total and Annular Eclipses of the Sun." *Memoirs of the Royal Astronomical Society* 10 (1838): 1–42.

Barton, Ruth. "'Men of Science': Language, Identity and Professionalization in the Mid-Victorian Scientific Community." *History of Science* 41, no. 1 (March 2003): 73–119.

Baum, Richard, and William Sheehan. *In Search of Planet Vulcan: The Ghost in Newton's Clockwork Universe*. Plenum Press, 1997.

Bellon, Richard, "Joseph Dalton Hooker's Ideals for a Professional Man of Science." *Journal of the History of Biology* 34, no. 1 (Spring 2001): 51–82.

Bellon, Richard. "Science at the Crystal Focus of the World." In Fyfe and Lightman, *Science in the Marketplace*, 301–35.

Bennett, Scott. "The Editorial Character and Readership of *The Penny Magazine*: An Analysis." *Victorian Periodical Review* 17, no. 4 (Winter 1984): 126–41.

Bennett, Scott. "Revolutions in Thought: Serial Publication and the Mass Market for Reading." In *The Victorian Periodical Press*, edited by Joanne Shattock and Michael Wolff, 225–57. Leicester University Press, 1982.

Bennett, Susan. *Theatre Audiences: A Theory of Production and Reception*, 2nd ed. Routledge, 1997.

Berkowitz, Carin, and Bernard Lightman, eds. *Science Museums in Transition:*

BIBLIOGRAPHY

Cultures of Display in Nineteenth-Century Britain and America. University of Pittsburgh Press, 2017.

Berman, Morris. *Social Change and Scientific Organization: The Royal Institution, 1799–1844*. Heinemann Education, 1978.

"Biographical Sketch of William Walker, ESQ., Lecturer on the Eidouranion, or Transparent Orrery." *Monthly Mirror* 7 (1799): 9–12.

Bird, Wendy. "Enlightenment and Entertainment: The Magic Lantern in Late 18th- and Early 19th-century Madrid." In Crangle et al., *Realms of Light*, 90–91.

Blunt, Blanche Elizabeth. *Uranographia: The Beauty of the Heavens, A Graphic Display of the Astronomical Phenomena of the Universe*. London: Simpkin & Marshall; J. Williams, Charles Street, Soho; G. Hebert, Cheapside; and all booksellers, 1836.

Blunt, Charles F. *The Beauty of the Heavens: A Pictorial Display of the Astronomical Phenomena of the Universe*, 2nd ed. London: Tilt and Bogue, 1842.

Bonnycastle, John. *An Introduction to Astronomy, in a Series of Letters from a Preceptor to His Pupil*. London: Printed for J. Johnson, 1786.

Booth, Michael R. *Theatre in the Victorian Age*. Cambridge University Press, 1991.

Bowler, Peter. *Evolution: The History of An Idea*, 25th anniversary ed. University of California Press, 2009.

Bowler, Peter, and Iwan R. Morus. *Making Modern Science: A Historical Survey*, 2nd ed. University of Chicago Press, 2020.

British Museum. "Drawing." Inventory no. 1862,0614.689. https://www.britishmuseum.org/collection/object/P_1862-0614-689.

Broks, Peter. *Understanding Popular Science*. Open University Press, 2006.

Brooke, John H. "Natural Theology and the Plurality of Worlds: Observations on the Brewster-Whewell Debate." *Annals of Science* 34, no. 3 (1977): 221–86.

Brooke, John H. *Science and Religion: Some Historical Perspectives*. Cambridge University Press, 1991.

Brooke, John H., and Geoffrey Cantor. *Reconstructing Nature: The Engagement of Science and Religion*. Oxford University Press, 1998.

[Brougham, Henry.] *A Discourse of the Objects, Advantages, and Pleasures of Science*. London: Baldwin, and Joy, 1827.

Brück, Mary T. *Agnes Mary Clerke and the Rise of Astrophysics*. Cambridge University Press, 2002.

BIBLIOGRAPHY

Brück, Mary T. *Women in Early British and Irish Astronomy: Stars and Satellites.* Springer, jointly published with the Royal Astronomical Society, 2009.

Brush, Stephen G. "The Nebular Hypothesis and the Evolutionary Worldview." *History of Science* 25, no. 3 (1987): 245–78.

Buick, Tony. *Orreries, Clocks, and London Society: the Evolution of Astronomical Instruments and Their Makers*, 2nd ed. Springer, 2020.

Burke, Edmund. *A Philosophical Enquiry into the Origin of Our Ideas of the Sublime and Beautiful.* London: R. and J. Dodsley, 1757.

Bury, John B. *The Idea of Progress: An Inquiry of Its Origin and Growth.* Macmillan, 1920.

Busby, Charles A. "Papers in Mechanics: No. IX, Hydraulic Orrery." *Transactions of the Society, Instituted at London, for the Encouragement of Arts, Manufactures, and Commerce* 40 (1822): 98–104.

Bush, Martin. "Again with Feeling: Modes of Visual Representation in Popular Astronomy in the Mid-Nineteenth Century." *Notes and Records: The Royal Society Journal of the History of Science* 76, no. 3 (September 2022): 485–506.

Bush, Martin. "The Astronomical Lantern Slide Set and the Eidouranion in Australia." *Early Popular Visual Culture* 17, no. 1 (2019): 9–33.

Bush, Martin, and Hsiang-Fu Huang. "New Stars on Screen: Continuities in the Development of Astronomical Lantern Slides in Britain, 1790–1850." *Centaurus*: forthcoming.

Butterworth, Mark. "Astronomical Lantern Slides." *New Magic Lantern Journal* 10, no. 4 (Autumn 2008): 65–68.

Butterworth, Mark. "Astronomy and the Magic Lantern." *Culture and Cosmos: A Journal of the History of Astrology and Cultural Astronomy* 8, no. 1 and 2 (2004): 15–32.

Butterworth, Mark. "A Lantern Tour of Star-Land: The Astronomer Robert Ball and His Magic Lantern Lectures." In Crangle et al., *Realms of Light*, 162–73.

Cane, R. F. "John H. Pepper: Analyst and Rainmaker." *Journal of the Royal Historical Society of Queensland* 9 (1975): 116–28.

Cantor, Geoffrey, ed. *The Great Exhibition: A Documentary History*, vol. 3. Pickering and Chatto, 2013.

Cantor, Geoffrey. *Religion and the Great Exhibition of 1851.* Oxford University Press, 2011.

Cantor, Geoffrey. "Thomas Young's Lectures at the Royal Institution." *Notes and Records of the Royal Society of London* 25, no. 1 (June 1970): 87–112.

BIBLIOGRAPHY

Cardwell, D. S. L. *The Organisation of Science in England*, rev. ed. Heinemann Educational, 1972.

Carlile, Richard. *An Address to Men of Science*. London: Printed and published by R. Carlile, 1821.

Carlile, Richard. "Astronomers and Astrologers." *Republican* 11 (January 21, 1825): 89–95.

Carlyle, E. I. "Walker, Adam (1730/31–1821)," revised by Anita McConnell. *Oxford Dictionary of National Biography*. https://doi.org/10.1093/ref:odnb/28466.

"The Central Sun." *Chambers's Edinburgh Journal* 7, no. 163 (February 13, 1847): 99–101 (BP PQ).

Chapman, Allan. "Science and the Public Good: George Biddell Airy (1801–1892) and the Concept of a Scientific Civil Servant." In *Science, Politics and the Public Good: Essays in Honour of Margaret Gowing*, edited by Nicolaas Rupke, 36–62. Macmillan, 1988.

Chapman, Allan. *The Victorian Amateur Astronomer: Independent Astronomical Research in Britain 1820–1920*. Praxis, 1998.

"Chelmsford Mechanics' Institute." *Essex Standard*, April 28, 1837 (BLN GPS).

"City Public Lectures." *Jackson's Oxford Journal*, December 27, 1851 (BLN GPS).

Clark, Raymond. "William Berry and His History of Guernsey." *Report and Transactions of la Société Guernesiaise* 22 (1987): 258–73.

Clarke, David. *Reflections on the Astronomy of Glasgow: A Story of Some 500 Years*. Edinburgh University Press, 2013.

Clerke, A. M. "Grant, Robert (1814–92)," revised by David Gavine. *Oxford Dictionary of National Biography*. https://doi.org/10.1093/ref:odnb/11285.

Clerke, A. M. "Pearson, William (1767–1847)," revised by David Philip Miller. *Oxford Dictionary of National Biography*. https://doi.org/10.1093/ref:odnb/21725.

Clifton, Gloria. *Directory of British Scientific Instrument Makers 1550–1851*. Zwemmer, in association with the National Maritime Museum, 1995.

Cockburn, J. S., H. P. F. King, and K. G. T. McDonnell. "Schools: Latymer and Godolphin Schools." In *A History of the County of Middlesex: Volume 1: Physique, Archaeology, Domesday, Ecclesiastical Organization, The Jews, Religious Houses, Education of Working Classes to 1870, Private Education from Sixteenth Century*, edited by J. S. Cockburn, H. P. F. King, and K. G. T. McDonnell, 305–6. Victoria County History: Middlesex, 1969.

BIBLIOGRAPHY

Collection of Historical Scientific Instruments, Harvard University. "Grand Orrery." Inventory no. 0004. https://chsi.emuseum.com/objects/3064/.

"Colosseum." *Morning Chronicle*, January 5, 1857.

Cooter, Roger, and Steven Pumfrey. "Separate Spheres and Public Places: Reflections on the History of Science Popularization and Science in Popular Culture." *History of Science* 32, no. 3 (1994): 237–67.

Corsi, Pietro. "Powell, Baden (1796–1860)." *Oxford Dictionary of National Biography*. https://doi.org/10.1093/ref:odnb/22642.

"Country News." *Gentleman's Magazine* 87 (January 1817): 76–78.

Crangle, Richard, Mervyn Heard, and Ine van Dooren, eds. *Realms of Light: Uses and Perceptions of the Magic Lantern from the 17th to the 21st Century*. Magic Lantern Society, 2005.

Croarken, Mary. "Astronomical Labourers: Maskelyne's Assistants at the Royal Observatory, Greenwich, 1765–1811." *Notes and Records of the Royal Society of London* 57, no. 3 (September 2003): 285–98.

Crop the Conjuror. "Hints to Lecturers." *The Satirist, or, Monthly Meteor* 3 (December 1808): 508–13 (BP PQ).

Crowe, Michael J. "Astronomy and Religion (1780–1915): Four Case Studies Involving Ideas of Extraterrestrial Life." *Osiris* 16 (2001): 209–26.

Crowe, Michael J. *The Extraterrestrial Life Debate 1750–1900: The Idea of a Plurality of Worlds from Kant to Lowell*. Cambridge University Press, 1986.

Deane, William. *The Description of the Copernican System, with the Theory of the Planets*. London, 1738.

"Death of Mr. George Bartley." *Era*, July 25, 1858.

"The Deformito-Mania." *Punch* 13 (September 4, 1847): 90.

Desaguliers, John T. *A Course of Experimental Philosophy*, vol. 1. London, 1734.

Desaguliers, John T. *Physico-Mechanical Lectures, or An Account of What Is Explained and Demonstrated in the Course of Mechanical and Experimental Philosophy*. London, 1717.

Desmond, Adrian. *The Politics of Evolution: Morphology, Medicine, and Reform in Radical London*. University of Chicago Press, 1989.

Desmond, Adrian. "Redefining the X Axis: 'Professionals,' 'Amateurs' and the Making of Mid-Victorian Biology; A Progress Report." *Journal of History of Biology* 34, no. 1 (Spring 2001): 3–50.

Desmond, Adrian. *Reign of the Beast: The Atheist World of W. D. Saull and His Museum of Evolution*. Open Book, 2024. https://doi.org/10.11647/OBP.0393.

"The Destruction of Saville House." *Leeds Mercury,* March 3, 1865 (BLN GPS).

Dewhirst, D. W. "The Greenwich-Cambridge Axis." *Vistas in Astronomy* 20, no. 1 (1976): 109–11.

Dick, Stephen J. *Plurality of Worlds: The Origin of the Extraterrestrial Life Debate from Democritus to Kant.* Cambridge University Press, 1982.

[Dick, Thomas.] *The Solar System,* Part I. London: Religious Tract Society, 1799.

Dickens, Charles. "The Uncommercial Traveller." *All the Year Round* 9 (June 6, 1863): 348–52.

Dierig, Sven, Jens Lachmund, and J. Andrew Mendelsohn. "Introduction: Toward an Urban History of Science." *Osiris,* 2nd series, 18, no. 1 (2003): 1–19.

"The Discoverer of the New Planet." *Bradford Observer,* February 2, 1860.

"The Discoverer of the New Planet." *Daily News,* January 28, 1860.

"The Discoverer of the New Planet." *Jackson's Oxford Journal,* February 4, 1860.

Donohue, Joseph, ed. *The Cambridge History of British Theatre, Volume 2: 1660 to 1895.* Cambridge University Press, 2004.

"The Drama and Public Amusements: Royal Polytechnic Institution, Lectures on Astronomy." *Critic* 5 (March 6, 1847): 193.

Drayton, Richard. *Nature's Government: Science, Imperial Britain, and the "Improvement" of the World.* Yale University Press, 2000.

"Easter Amusements, &c." *Illustrated London News* 30, no. 853 (April 11, 1857): 337 (ILNHA GPS).

"Eclipses." *Aberdeen Journal,* January 2, 1861.

Edinburgh Society of Model Engineers. "A Replica of the Orrery Built by the Revd. William Pearson." https://edinburgh-sme.org.uk/a-replica-of-the-orrery-built-by-the-revd-william-pearson/.

Elliott, Paul A. "The Birth of Public Science in the English Provinces: Natural Philosophy in Derby, c. 1690–1760." *Annals of Science* 57, no. 1 (2000): 61–100.

Elliott, Paul A. *The Derby Philosophers: Science and Culture in British Urban Society, 1700–1850.* Manchester University Press, 2009.

[Ellis, Robert, ed.] *Official Catalogue of the Great Exhibition of the Works of Industry of All Nations, 1851,* supplementary volume. London: Spicer Brothers, 1851.

Endersby, Jim. *Imperial Nature: Joseph Hooker and the Practices of Victorian Science.* University of Chicago Press, 2008.

"Faith in Astronomy." *London Saturday Journal* 1, no. 22 (June 1, 1839): 350–51.

Fara, Patricia. "Desaguliers, John Theophilus (1683–1744)." *Oxford Dictionary of National Biography*. https://doi.org/10.1093/ref:odnb/7539.

Faraday, Michael. "On Mr. Warren de la Rue's Photographic Eclipse Results." *Notices of the Proceedings at the Meetings of Members of the Royal Institution of Great Britain* 3 (May 3, 1861 [1858–62]): 362–66.

Ferguson, James. *Astronomy Explained upon Sir Isaac Newton's Principles: And Made Easy to Those Who Have Not Studied Mathematics*. London: Printed for, and sold by the author, 1756.

"Fine Arts." *Metropolitan Magazine* 12, no. 47 (March 1835): 85 (BP PQ).

Finnegan, Diarmid A. "Placing Science in an Age of Oratory: Spaces of Scientific Speech in Mid-Victorian Edinburgh." In Livingstone and Withers, *Geographies of Nineteenth-Century Science*, 153–77.

Finnegan, Diarmid A. "The Spatial Turn: Geographical Approaches in the History of Science." *Journal of the History of Biology* 41, no. 2 (Summer 2008): 369–88.

Finnegan, Diarmid A. *The Voice of Science: British Scientists on the Lecture Circuit in Gilded Age America*. University of Pittsburgh Press, 2021.

Flanders, Judith. *The Victorian City: Everyday Life in Dickens' London*. Atlantic Books, 2012.

Forgan, Sophie. "The Architecture of Display: Museums, Universities and Objects in Nineteenth-Century Britain." *History of Science* 32, no. 2 (June 1994): 139–62.

Forgan, Sophie. "Building the Museum: Knowledge, Conflict, and the Power of Place." *Isis* 96, no. 4 (December 2005): 572–85.

Forgan, Sophie. "'A National Treasure House of a Unique Kind' (W. L. Bragg): Some Reflections on Two Hundred Years of Institutional History." In James, *"The Common Purposes of Life,"* 17–42.

Forgan, Sophie. "The Royal Institution of Great Britain, 1840–1873." PhD diss., University of London, 1977.

Fortier, Mark. *Theory/Theatre: An Introduction*. Routledge, 1997.

Foulkes, Richard. *Church and Stage in Victorian England*. Cambridge University Press, 1997.

Fox, Celina, ed. *London: World City 1800–1840*. Yale University Press, in association with the Museum of London, 1992.

[Francis, George W.] "Astronomical Illustrations." *Magazine of Science* 1 (1840): 329–31.

BIBLIOGRAPHY

[Francis, George W.] "Astronomical Illustrations." *Magazine of Science* 2 (1841): 1–3.

Frayling, Christopher. *Mad, Bad, and Dangerous? The Scientist and the Cinema.* Reaktion, 2005.

Freeman, Michael. *Railways and Victorian Imagination.* Yale University Press, 1999.

Fyfe, Aileen. "Publishing and the Classics: Paley's *Natural Theology* and the Nineteenth-Century Scientific Canon." *Studies in History and Philosophy of Science Part A* 33, no. 4 (December 2002): 729–51.

Fyfe, Aileen. *Science and Salvation: Evangelical Popular Science Publishing in Victorian Britain.* University of Chicago Press, 2004.

Fyfe, Aileen. *Steam-Powered Knowledge: William Chambers and the Business of Publishing, 1820–1860.* University of Chicago Press, 2012.

Fyfe, Aileen, and Bernard Lightman. "Science in the Marketplace: An Introduction." In Fyfe and Lightman, *Science in the Marketplace,* 1–19.

Fyfe, Aileen, and Bernard Lightman, eds. *Science in the Marketplace: Nineteenth-Century Sites and Experiences.* University of Chicago Press, 2007.

Garvey, M. A. *The Silent Revolution: Or the Future Effects of Steam and Electricity upon the Condition of Mankind.* London: William and Frederick G. Cash, 1852.

"Gas as Fuel." *Bradford Observer,* November 1, 1873.

Gascoigne, John. "From Bentley to the Victorians: The Rise and Fall of British Newtonian Theology." *Science in Context* 2, no. 2 (Autumn 1988): 219–56.

Gates, Barbara T. *Kindred Nature: Victorian and Edwardian Women Embrace the Living World.* University of Chicago Press, 1998.

"George III: A Royal Passion for Science." Science Museum, August 13, 2019. https://www.sciencemuseum.org.uk/objects-and-stories/george-iii-royal-passion-science.

Gieryn, Thomas. "City as Truth-Spots: Laboratories and Field-Sites in Urban Studies." *Social Studies of Science* 36, no. 1 (February 2006): 5–38.

Gieryn, Thomas. "Three Truth-Spots." *Journal of the History of Behavioral Sciences* 38, no. 2 (April 2002): 113–32.

Gilfillan, George. *Young's Night Thoughts: With Life, Critical Dissertation, and Explanatory Notes.* Edinburgh: James Nichol, 1853. Project Gutenberg.

Godwin, William. *Thoughts on Man: His Nature, Production and Discoveries Interspersed Some Particulars Respecting the Author.* London: Effingham Wilson, Royal Exchange, 1831. Project Gutenberg.

Golinski, Jan. "A Noble Spectacle: Phosphorus and the Public Cultures of Science in the Early Royal Society." *Isis* 80, no. 1 (March 1989): 11–39.

Golinski, Jan. *Science as Public Culture: Chemistry and Enlightenment in Britain, 1760–1820*. Cambridge University Press, 1992.

Golinski, Jan. "Sublime Astronomy: The Eidouranion of Adam Walker and His Sons." *Huntington Library Quarterly* 80, no. 1 (Spring 2017): 135–57.

Gosse, Edmund. *Father and Son: A Study of Two Temperaments*, edited by Peter Abbs. Penguin Books, 1983.

Gouyon, Jean-Baptiste. "1985, Scientists Can't Do Science Alone, They Need Publics." *Public Understanding of Science* 25, no. 6 (August 2016): 754–57.

Grant, James. *The Great Metropolis*, 3rd ed., vol. 1. New York: Saunders and Otley, 1837.

Gregory, Jane, and Steve Miller. *Science in Public: Communication, Culture, and Credibility*. Basic Books, 1998.

Grogans, R. M. "Fulton's Orrery." *Journal of the British Astronomical Association* 88, no. 3 (1978): 277–80.

Gross, Alan G. *The Scientific Sublime: Popular Science Unravels the Mystery of the Universe*. Oxford University Press, 2018.

Gurman, S. J., and S. R. Harratt. "Revd Dr William Pearson (1767–1847): A Founder of the Royal Astronomical Society." *Quarterly Journal of the Royal Astronomical Society* 35, no. 3 (September 1994): 271–92.

Guthke, Karl S. "Nightmare and Utopia: Extraterrestrial Worlds from Galileo to Gothe." *Early Science and Medicine* 8, no. 3 (2003): 173–95.

Harvard University, David P. Wheatland, and Barbara Carson. *The Apparatus of Science at Harvard, 1765–1800*. Harvard University Press, 1968.

"Harwich: Lecture on Astronomy." *Essex Standard*, March 16, 1866 (BLN GPS).

Haynes, Roslynn D. *From Faust to Strangelove: Representations of the Scientist in Western Literature*. Johns Hopkins University Press, 1994.

Hays, Jo N. "The London Lecturing Empire, 1800–50." In *Metropolis and Province: Science in British Culture, 1780–1850*, edited by Ian Inkster and Jack Morrell, 91–119. Hutchinson, 1983.

Hays, Jo N. "Science in the City: The London Institution, 1819–40." *British Journal for the History of Science* 7, no. 2 (July 1974): 146–62.

Heard, Mervyn. *Phantasmagoria: The Secret Life of the Magic Lantern*. Projection Box, 2006.

Henderson, Ebenezer. *A Treatise on Astronomy, Displaying the Arithmetical Architecture of the Solar System*, 2nd ed. London: N. H. Cotes, 1843.

BIBLIOGRAPHY

Henderson, Heather. *The Victorian Self: Autobiography and Biblical Narrative.* Cornell University Press, 1989.

Herrmann, Dieter B. *The History of Astronomy from Herschel to Hertzsprung.* Translated and edited by Kevin Krisciunas. Cambridge University Press, 1984.

Herschel, John F. W. "Address." In *Report of the Fifteenth Meeting of the British Association for the Advancement of Science; Held at Cambridge in June 1845,* xxxvii–xliv. London: John Murray, 1846.

Herschel, John F. W. *Essays from Edinburgh and London Quarterly Reviews, with Addresses and Other Pieces.* London: Longman, 1857.

Herschel, John F. W. *A Treatise on Astronomy.* London: Longman, Rees, Orme, Brown, Green, & Longman, and John Taylor, 1834.

Higgitt, Rebekah. *Recreating Newton: Newtonian Biography and the Making of Nineteenth-Century History of Science.* Pickering and Chatto, 2007.

Higgitt, Rebekah, and Charles W. J. Withers. "Science and Sociability: Women as Audience at the British Association for the Advancement of Science, 1831–1901." *Isis* 99, no. 1 (March 2008): 1–27.

"History of the Christmas Lectures." Royal Institution of Great Britain. https://www.rigb.org/christmas-lectures/history-christmas-lectures.

History of Science Museum, University of Oxford. "Astronomical Demonstration Device, English, 1817." Inventory no. 52003. https://www.hsm.ox.ac.uk/collections-online#/item/hsm-catalogue-2761.

Hodgson, James, and Francis Hauksbee. *An Account of Hydrostatical and Pneumatical Experiments.* [London], 1715.

Hollingshead, John. "At the Play." *Cornhill Magazine* 5 (January 1862): 84–92 (BP PQ).

Holmes, Richard. *The Age of Wonder: How the Romantic Generation Discovered the Beauty and Terror of Science.* Harper Collins, 2008.

Hone, William. *The Real or Constitutional House That Jack Built.* London: J. Asperne, 1819.

Hood, Thomas. "Love and Lunacy." *Comic Annual* 7 (1836): 33–82 (BP PQ).

Hoskin, Michael. *Discoverers of the Universe: William and Caroline Herschel.* Princeton University Press, 2011.

Hoskin, Michael. "John Herschel's Cosmology." *Journal for the History of Astronomy* 18, no. 1 (February 1987): 1–34.

Huang, Hsiang-Fu. "From Grub Street to the Colony: George William Francis and an Early Victorian Scientific Career." *Notes and Records: The Royal Society Journal of the History of Science* 76, no. 1 (March 2022): 181–208.

BIBLIOGRAPHY

Huang, Hsiang-Fu, ed. *Ouranologia: An Annotated Edition of a Lenten Lecture on Astronomy with Critical Introduction.* UCL Department of Science and Technology Studies, 2015. https://www.ucl.ac.uk/sts/file/8359.

Hutchins, Roger. *British University Observatories 1772–1939.* Routledge, 2016.

Hunt, Frederick K. "The Planet Watchers of Greenwich." *Household Words* 9 (May 25, 1850): 200–204.

Hunt, Verity. "Raising a Modern Ghost: The Magic Lantern and the Persistence of Wonder in the Victorian Education of the Senses." *Romanticism and Victorianism on the Net,* no. 52 (November 2008). https://doi.org/10.7202/019806ar.

"The Hydroelectric Machine." *Illustrated Polytechnic Review* 2 (1843): 162–63.

"Hydro-Electric Machine." *Times,* September 15, 1843, 7.

"Ilford and Barking." *Essex Standard,* January 30, 1856 (BLN GPS).

Inkster, Ian. "Advocates and Audience: Aspects of Popular Astronomy in England, 1750–1850." *The Journal of the British Astronomical Association* 92 (1982): 119–23.

Inkster, Ian. "Science and the Mechanics' Institutes, 1820–1850: The Case of Sheffield." *Annals of Science* 32, no. 5 (1975): 451–74.

"Inspiring Minds Since 1824." Bath Royal Literary and Science Institution. Accessed March 5, 2024, https://www.brlsi.org/brlsi200/.

"An Interesting Astronomical Episode: A New Planet Within Mercury." *Glasgow Herald,* February 9, 1860 (BLN GPS).

Jackson, Lee. "Adelaide Gallery." *Dictionary of Victorian London.* https://www.victorianlondon.org/entertainment/adelaidegallery.htm.

Jackson, Russell, ed. *Victorian Theatre: The Theatre in Its Time.* New Amsterdam Books, 1994.

Jacob, Margaret C., and Larry Stewart. *Practical Matter: Newton's Science in the Service of Industry and Empire, 1687–1851.* Harvard University Press, 2004.

James, Frank A. J. L., ed. *Christmas at the Royal Institution: An Anthology of Lectures by M. Faraday, J. Tyndall, R. S. Ball, S. P. Thompson, E. R. Lankester, W. H. Bragg, W. L. Bragg, R. L. Gregory, and I. Stewart.* World Scientific, 2007.

James, Frank A. J. L., ed. *"The Common Purposes of Life": Science and Society at the Royal Institution of Great Britain.* Ashgate, 2002.

James, Frank A. J. L., ed. *The Correspondence of Michael Faraday,* vol. 1. Institution of Electrical Engineers, 1991.

James, Frank A. J. L. "Introduction." In James, *"Common Purposes of Life,"* 1–16.

James, Frank A. J. L. "Reporting Royal Institution Lectures, 1826–1867." In *Science Serialized: Representations of the Sciences in Nineteenth-Century Periodicals*, edited by Geoffrey Cantor and Sally Shuttleworth, 67–79. MIT Press, 2004.

James, Frank A. J. L. "Running the Royal Institution: Faraday as an Administrator." In James, *"The Common Purposes of Life,"* 119–46.

James, Frank A. J. L., and Anthony Peers. "Constructing Space for Science at the Royal Institution of Great Britain." *Physics in Perspective* 9 (2007): 130–85.

Jerrold, Walter. *Thomas Hood: His Life and Times*. Alston Rivers, 1907.

Jones, Peter M. *Industrial Enlightenment: Science, Technology and Culture in Birmingham and the West Midlands, 1760–1820*. Manchester University Press, 2008.

Jones, Roger. "Sir Robert Ball: Victorian Astronomer and Lecturer par Excellence." *Antiquarian Astronomer*, no. 2 (December 2005): 27–36.

[Jones, William.] *A Catalogue of Optical, Mathematical, and Philosophical Instruments, Made and Sold by W. and S. Jones*. London, 1797.

[Jones, William.] *A Catalogue of Optical, Mathematical, and Philosophical Instruments, Made and Sold by W. and S. Jones*. London: W. Glendinning, 1814.

Jones, William. *The Description and Use of a New Portable Orrery*, 4th ed. London: Printed for, and sold by, W. and S. Jones, 1794.

"Joseph Wright of Derby." Derby Museum and Art Gallery. https://derbymuseums.org/collection/joseph-wright-of-derby/.

J. W. S. "The Wonders of the Heavens." *Youth's Magazine*, 3rd series, 6 (August 1833): 265–67 (SciPer).

Keene, Melanie. "Familiar Science in Nineteenth-Century Britain." *History of Science* 52, no. 1 (March 2014): 53–71.

Keene, Melanie. "Playing Among the Stars: *Science in Sport, or the Pleasures of Astronomy* (1804)." *History of Education* 40, no. 4 (September 2011): 521–42.

Kelly, Thomas. *A History of Adult Education in Great Britain*, 2nd ed. Liverpool University Press, 1970.

King, Henry C., and John R. Millburn. *Geared to the Stars: The Evolution of Planetariums, Orreries and Astronomical Clocks*. University of Toronto Press, 1978.

"King's Theatre." *Theatrical Observer*, no. 4428 (February 25, 1836): [1] (BP PQ).

Kirwan, James. *Sublimity*. Routledge, 2005.

Kitchener, William. *The Economy of the Eyes*. Boston: Wells and Lilly, 1824.

Knight, David. *The Making of Modern Science: Science, Technology, Medicine and Modernity: 1789–1914*. Polity, 2009.

Knight, David. *Public Understanding of Science: A History of Communicating Scientific Ideas*. Routledge, 2006.

Knight, Joseph. "Bartley, George (bap. 1784, d. 1858)," revised by Katharine Cockin. *Oxford Dictionary of National Biography*. https://doi.org/10.1093/ref:odnb/1589.

Kohlstedt, Sally G. "Parlors, Primers, and Public Schooling: Education for Science in Nineteenth-Century America." *Isis* 81, no. 3 (September 1990): 424–45.

Kurzer, Frederick. "Chemistry and Chemists at the London Institution 1807–1912." *Annals of Science* 58, no. 2 (2001): 163–201.

Kurzer, Frederick. "A History of the Surrey Institution." *Annals of Science* 57, no. 2 (2000): 109–41.

Lach-Szyrma, Krystyn. *London Observed: A Polish Philosopher at Large, 1820–24*. Translated by Margozata Machnice and Agnieszka Kiersztejn, edited by Mona K. McLeod. Signal Books, 2009.

Lacy, William. *An Introduction to Astronomy, by Which the Knowledge of the Solar System Is Rendered Extremely Easy to Youth, and Those Who Have Not Studied Mathematics*. London, 1777.

Lamb, Charles. "The Old and the New Schoolmaster." *London Magazine* 3 (May 1821): 492–97.

Langford, Paul. *A Polite and Commercial People: England 1727–1783*. Oxford: Oxford University Press, 1989.

"The Late Solar Eclipse." *Illustrated London News* 1, no. 12 (July 30, 1842): 180 (ILNHA GPS).

"The Late Solar Eclipse." *Morning Post*, July 25, 1842 (BLN GPS).

"Lecture by the Senior Wrangler." *Essex Standard*, February 17, 1865 (BLN GPS).

"Lecture in Lent." *Punch* 60 (1871): 105.

"Lecture on Astronomy." *Leeds Mercury*, May 8, 1852 (BLN GPS).

"A Lecturer of the Old School." *Leisure Hour* 2, no. 95 (October 20, 1853): 676–78 (BP PQ).

"Lectures and the Lecturers at the Mechanics' Institution." *Liverpool Mercury*, November 14, 1845 (BLN GPS).

BIBLIOGRAPHY

"Lectures on Astronomy." *Belfast News-Letter*, November 18, 1842 (BLN GPS).

"Lectures on Astronomy." *Hampshire Advertiser*, November 2, 1844 (BLN GPS).

"Lectures on Astronomy, by John Wallis, Esq." *Examiner*, September 30, 1827, 623 (BP PQ).

Lee, Hermione. "Writing Victorian Lives and Victorian Life-Writing: Gosse's 'Father and Son' Revisited." *Journal of Victorian Culture* 8, no. 1 (January 2003): 108–18.

Levenson, Thomas. *The Hunt for Vulcan... And How Albert Einstein Destroyed a Planet, Discovered Relativity, and Deciphered the Universe*. Random House, 2015.

Levitt, Theresa. "'I Thought This Might be of Interest...': The Observatory as Public Enterprise." In Aubin et al., *Heavens on Earth*, 285–304.

Lightman, Bernard. "Lecturing in the Spatial Economy of Science." In Fyfe and Lightman, *Science in the Marketplace*, 97–132.

Lightman, Bernard. "Pepper, John Henry (1821-1900)." Oxford Dictionary of National Biography. https://doi.org/10.1093/ref:odnb/21898.

Lightman, Bernard. "Refashioning the Spaces of London Science: Elite Epistemes in the Nineteenth Century." In Livingstone and Withers, *Geographies of Nineteenth-Century Science*, 25–50.

Lightman, Bernard. "Science in Regent's Park: The Colosseum." In Berkowitz and Lightman, *Science Museums in Transition*, 13–35.

Lightman, Bernard. "Spectacle in Leicester Square: James Wyld's Great Globe, 1851–1861." In *Popular Exhibitions, Science and Showmanship, 1840–1910*, edited by Joe Kember, John Plunkett, and Jill A. Sullivan, 19–39. Pickering and Chatto, 2012.

Lightman, Bernard. *Victorian Popularizers of Science: Designing Nature for New Audiences*. University of Chicago Press, 2007.

Lightman, Bernard, and Michael S. Reidy, eds. *The Age of Scientific Naturalism: Tyndall and His Contemporaries*. Pickering and Chatto, 2014.

"Literary and Art Gossip." *Leeds Mercury*, August 6, 1879 (BLN GPS).

Littmann, Mark, Fred Espenak, and Ken Willcox. *Totality: Eclipses of the Sun*. Oxford University Press, 2008.

Livingstone, David N. *Putting Science in Its Place: Geographies of Scientific Knowledge*. University of Chicago Press, 2003.

Livingstone, David N. "Science, Site and Speech: Scientific Knowledge and the Spaces of Rhetoric." *History of the Human Sciences* 20, no. 2 (May 2007): 71–98.

BIBLIOGRAPHY

Livingstone, David N., and Charles W. J. Withers, eds. *Geographies of Nineteenth-Century Science*. University of Chicago Press, 2011.

Lloyd, Harriet O. "Rulers of Opinion: Women at the Royal Institution of Great Britain, 1799–1812." PhD diss., University College London, 2018.

Locke, Simon. "Golem Science and the Public Understanding of Science: From Deficit to Dilemma." *Public Understanding of Science* 8, no. 2 (April 1999): 75–92.

"The London Exhibitions, &c.: Colosseum." *Era*, January 3, 1864.

"London Mechanics' Institution: Lectures on Astronomy." *Examiner*, March 19, 1826 (BLN GPS).

A Lover of Truth. "Copy of a Letter Sent to Mr. Rogers, Itinerant Lecturer on Astronomy." *Republican* 11 (January 21, 1825): 88–89.

Lubbock, Constance A., ed. *The Herschel Chronicle: The Life-Story of William Herschel and His Sister Caroline Herschel*. Macmillan, 1933. Cambridge University Press, 2013.

Mackintosh, Iain. "Departing Glories of the British Theatre: Setting Suns over a Neoclassical Landscape." In Fox, *London*, 199–208.

Mackintosh, Iain. "Pit, Boxes and Gallery." In Fox, *London*, 553.

The Magic Lantern Society. "A Brief History of the Magic Lantern." https://www.magiclantern.org.uk/history/.

Mander, Raymond, and Joe Mitchenson. *The Theatres of London*. Rupert Hart-Davis, 1961.

Marché, Jordan D, II. *Theaters of Time and Space: American Planetaria, 1930–1970*. Rutgers University Press, 2005.

Martin, Benjamin. *The Description and Use of an Orrery of a New Construction*. London: Printed for, and sold by the author, 1771.

Martin, Benjamin. *The Description and Use of Both the Globes, Armillary Sphere and Orrery*. London, 1762.

Martin, Benjamin. *The Young Gentleman and Lady's Philosophy, in a Continued Survey of the Works of Natural and Art; by way of Dialogue*, vol. 1. London, 1759.

Martin, Philip W. "Carlile, Richard (1790–1843)." *Oxford Dictionary of National Biography*. https://doi.org/10.1093/ref:odnb/4685.

McWilliam, Rohan. *London's West End: Creating the Pleasure District, 1800–1914*. Oxford University Press, 2020.

"Mechanics' Institution at Liverpool." *Chambers's Edinburgh Journal* 11, no. 570 (December 31, 1842): 396–97.

"The Mechanics' Institutions of This District." *Newcastle Magazine* 6, no. 7 (July 1827): 297–98 (BP PQ).

BIBLIOGRAPHY

"Meeting of the Royal Astronomical Society, Friday, 1880, May 14." *Observatory* 3, no. 38 (June 1, 1880): 439.

"Memoir of Stephen Kemble." *Oxberry's Dramatic Biography and Histrionic Anecdotes* 2 (April 1825): 5–11.

A Mere Phantom. *The Magic Lantern: How to Buy, and How to Use It. Also How to Raise a Ghost*. London: Houlston and Wright, 1866.

Metropolitan Magazine 34 (February 1834): 56.

Millburn, John R. "Benjamin Martin and the Development of the Orrery." *British Journal for the History of Science* 6, no. 4 (December 1973): 378–99.

Millburn, John R. "Martin, Benjamin (bap. 1705, d. 1782)." *Oxford Dictionary of National Biography*. https://doi.org/10.1093/ref:odnb/18175.

Millburn, John R., and Henry C. King. *Wheelwright of the Heavens: The Life and Work of James Ferguson, FRS*. Vade-Mecum Press, 1988.

Milner, Thomas. *The Gallery of Nature: A Pictorial and Descriptive Tour Through Creation*. London: Wm. S. Orr & Co., 1846.

"Miscellaneous: Astronomical Lectures." *Musical World* 22, no. 14 (April 3, 1847): 223–25 (BP PQ).

"Miscellaneous Exhibitions: The Colosseum." *Standard*, December 27, 1856 (BLN GPS).

"Miscellaneous Intelligence: A Supposed New Interior Planet." *Monthly Notices of the Royal Astronomical Society* 20, no. 3 (January 1860): 98–99.

"Miscellaneous Intelligence: Suspected Existence of a Zone of Asteroids Revolving Between Mercury and the Sun." *Monthly Notices of the Royal Astronomical Society* 20, no. 1 (November 1859): 24–26.

Moore, William P. "The Wonders of Astronomy." In *Lectures Delivered Before the Young Men's Christian Association, in the Rotundo, Dublin, 1854–5*, 1–46. Dublin: Samuel B. Oldham, 8, Suffolk-Street, 1855.

Mollan, Charles, ed. *William Parsons, 3rd Earl of Rosse: Astronomy and the Castle in Nineteenth-Century Ireland*. Manchester University Press, 2014.

Morrell, Jack. "Practical Chemistry in the University of Edinburgh, 1799–1843." *Ambix* 16, no. 1–2 (1969): 66–80.

Morrell, Jack. "Professionalisation." In *Companion to the History of Modern Science*, edited by R. C. Olby, G. N. Cantor, J. R. R. Christie, and M. J. S. Hodge, 980–89. Routledge, 1990.

Morrell, Jack, and Arnold Thackray. *Gentlemen of Science: Early Years of the British Association for the Advancement of Science*. Clarendon Press, 1981.

Morton, A. Q., and J. A. Wess. *Public and Private Science: The King George III Collection*. Oxford University Press, in association with the Science Museum, 1993.

Morus, Iwan R. "Currents from the Underworld: Electricity and the Technology of Display in Early Victorian England." *Isis* 84, no. 1 (March 1993): 50–69.

Morus, Iwan R. *Frankenstein's Children: Electricity, Exhibition and Experiment in Early-Nineteenth-Century London*. Princeton University Press, 1998.

Morus, Iwan R. "'More the Aspect of Magic Than Anything Natural': The Philosophy of Demonstration." In Fyfe and Lightman, *Science in the Marketplace*, 336–70.

Morus, Iwan R. "Replacing Victoria's Scientific Culture." *19: Interdisciplinary Studies in the Long Nineteenth Century* 2 (2006). https://doi.org/10.16995/ntn.441.

Morus, Iwan R. "Worlds of Wonder: Sensation and the Victorian Scientific Performance." *Isis* 101, no. 4 (December 2010): 806–16.

Morus, Iwan R., Simon Schaffer, and James Secord. "Scientific London." In Fox, *London*, 129–42.

"Mr. Adams's Orrery." *Standard*, March 27, 1861 (BLN GPS).

"Mr. Deane Walker." *Gentleman's Magazine* 2 (July 1865): 113.

"Musical and Dramatic Gossip." *Athenaeum*, no. 1693 (April 7, 1860): 481–82 (BP PQ).

Mussell, James. "Arthur Cowper Ranyard, *Knowledge* and the Reproduction of Astronomical Photographs in the Late Nineteenth-Century Periodical Press." *British Journal for the History of Science* 42, no. 3 (September 2009): 345–80.

Mussell, James. "Private Practice and Public Knowledge: Science, Professionalization and Gender in the Late Nineteenth Century." *Nineteenth-Century Gender Studies*, no. 5.2 (Summer 2009). https://www.ncgsjournal.com/issue52/mussell.html.

Nall, Joshua. *News from Mars: Mass Media and the Forging of a New Astronomy. 1860–1910*. University of Pittsburgh Press, 2019.

Neeley, Kathryn A. *Mary Somerville: Science, Illumination, and the Female Mind*. Cambridge University Press, 2001.

"New Scientific Establishment at Bath." *Quarterly Journal of Science, Literature and the Arts* 8, no. 15 (October 1819): 190–91.

"Newcastle upon Tyne and North of England Protestant Alliance." *Newcastle Courant*, March 19, 1852 (BLN GPS).

BIBLIOGRAPHY

Nichol, J. P. *Views of the Architecture of the Heavens*, 3rd ed. Edinburgh: William Tait, 1839.

"Nota Bene." *Punch* 7 (1844): 95.

"Notice." *Poor Man's Guardian*, December 28, 1833 (BLN GPS).

Numbers, Ronald. *Creation by Natural Law: Laplace's Nebular Hypothesis in American Thought*. University of Washington Press, 1977.

"Obituary." *Gentleman's Magazine* 39 (February 1853): 217.

"Obituary: Daniel Moore, Esq. F.R.S." *Gentleman's Magazine* 98 (April 1828): 377 (BP PQ).

"Obituary of Eminent Persons." *Gentleman's Magazine* 86 (April 1816): 374 (BP PQ).

O'Connor, Ralph. *The Earth on Show: Fossils and the Poetics of Popular Science, 1802–1856*. University of Chicago Press, 2007.

O'Connor, Ralph. "Reflections on Popular Science in Britain: Genres, Categories, and Historians." *Isis* 100, no. 2 (June 2009): 333–45.

Ogilvie, Marilyn B. "Obligatory Amateurs: Annie Maunder (1868–1947) and British Women Astronomers at the Dawn of Professional Astronomy." *British Journal for the History of Science* 33, no. 1 (March 2000): 67–84.

Ogilvie, Marilyn B. "Robert Chambers and the Nebular Hypothesis." *British Journal for the History of Science* 8, no. 3 (November 1975): 214–32.

Ogilvie, Marilyn B. "Obligatory Amateurs: Annie Maunder (1868–1947) and British Women Astronomers at the Dawn of Professional Astronomy." *The British Journal for the History of Science* 33, no. 1 (March 2000): 67–84.

Oxford English Dictionary, s.v. "career (n.)." https://doi.org/10.1093/OED/3914855094.

Oxford English Dictionary, s.v. "sublime (adj. & n.)." https://doi.org/10.1093/OED/1040109669.

Page, David. *The Earth's Crust: A Handy Outline of Geology*. Edinburgh: William P. Nimmo, 1864.

Paley, William. *Natural Theology, or Evidences of the Existence and Attributes of the Deity*, 12th ed. London: Printed for J. Faulder, 1809.

Pang, Alex Soojung-Kim. *Empire and the Sun: Victorian Solar Eclipse Expeditions*. Stanford University Press, 2002.

Park, Richard H. R., and Maureen Park. "Goya's Living Skeleton." *British Medical Journal* 303 (December 1991): 1594–96. https://doi.org/10.1136/bmj.303.6817.1594.

Parolin, Christina. *Radical Spaces: Venues of Popular Politics in London, 1790–c. 1845*. ANU E Press, 2010. http://doi.org/10.22459/RS.12.2010.

Pearson, William. "Planetary Machines." In *The Cyclopaedia*, vol. 27, edited by Abraham Rees. London: Longman, Hurst, Rees, Orme, & Brown, 1819.

Pearson, William. "Planetary Machines." In *The Edinburgh Encyclopaedia*, vol. 16, edited by David Brewster. Edinburgh: William Blackwood, 1830.

Perkins, Adam. "'Extraneous Government Business': The Astronomer Royal as Government Scientist: George Airy and His Work on the Commissions of State and Other Bodies, 1838–1880." *Journal of Astronomical History and Heritage* 4, no. 2 (December 2001): 143–54.

"Phrenological Society." *Preston Chronicles*, March 23, 1839 (BLN GPS).

Picard, Liza. *Dr. Johnson's London: Life in London 1740–1770*. Phoenix Press, 2001.

Picard, Liza. *Victorian London: The Life of a City 1840–1870*. Phoenix Press, 2006.

Pickstone, John. *Ways of Knowing: A New History of Science, Technology and Medicine*. Manchester University Press, 2000.

Podmore, J. B. *A Lecture on Astronomy, Delivered in the Town Hall, Birmingham, June 9th, 1857*. Birmingham: Cornish Brothers, the Journal Buildings, 1857.

Poole, John. "A Discourse of Bores." *New Monthly Magazine* 52 (1838): 396–403; 551–61.

Porter, Roy. "Gentlemen and Geology: The Emergence of a Scientific Career." *Historical Journal* 21, no. 4 (December 1978): 809–36.

Powell, Richard C. "Geographies of Science: Histories, Localities, Practices, Futures." *Progress in Human Geography* 31, no. 3 (June 2007): 309–30.

Powers, Jonathan. *The Philosopher Lecturing on the Orrery: The Rediscovery of the Life and Career of John Arden of Derby, Bath, and Beverley*. Quandary Books, 2019.

Priaulx Library. "March 1811: From the Newspapers." https://www.priaulxlibrary.co.uk/node/426.

"Proceedings of the Meeting." *Report of the Third Meeting of the British Association for the Advancement of Science; held at Cambridge in 1833*. London: John Murray, 1834.

"Proceedings of the Royal Institution." *Quarterly Journal of Science, Literature and the Arts* 21 (1826): 114–16.

"Provincial Amusements." *Saturday Review of Politics, Literature, Science and Art* 14, no. 363 (October 11, 1862): 441–42 (BP PQ).

BIBLIOGRAPHY

"Provincial Occurrences, in the Counties of England, Arranged Alphabetically: Berkshire." *New Monthly Magazine* 6, no. 36 (January 1817): 549–60.

"Public Amusements: The Christmas Holidays." *Lloyd's Weekly Newspaper*, December 25, 1864 (BLN GPS).

Pumfrey, Steven. "Idea Above His Station: A Social Study of Hooke's Curatorship of Experiments." *History of Science* 29, no. 1 (March 1991): 1–44.

"Punch's Almanack for 1855." *Punch* 28 (1855): [8].

"Punch's Almanack for 1866." *Punch* 50 (1866): [10].

"Punch's Information for the People, no. 1: Being a Very Familiar Treatise on Astronomy." *Punch* 1 (August 7, 1841): 41.

Ratcliff, Jessica. *The Transit of Venus Enterprise in Victorian Britain*. Pickering and Chatto, 2008.

"Recollections of the Rev. John Eyton, A.M., Formerly Vicar Wellington, Salop." *Wesleyan-Methodist Magazine* 3 (June 1847): 551–59 (BP PQ).

Reid, Francis. *"Two Systems of Astronomy* (1846): Plebian Resistance and Scriptural Astronomy." *British Journal for the History of Science* 38, no. 2 (June 2005): 161–77.

Report from the Select Committee on Dramatic Literature, with Minutes of Evidence. London: Ordered by the House of Commons to be printed, 1832.

Report from the Select Committee on Theatrical Licenses and Regulations: Together with the Proceedings of the Committee, Minutes of Evidence, Appendix, and Index. London: Communicated from the Commons to the Lords, 1866.

Report of the Third Meeting of the British Association for the Advancement of Science; held at Cambridge in 1833. London: John Murray, 1834.

Reports by the Juries on the Subjects in the Thirty Classes into Which the Exhibition Was Divided. London: Printed for the Royal Commission, by William Clowes & Sons, 1852.

Riskin, Jessica. "Amusing Physics." In *Science and Spectacle in the European Enlightenment*, edited by Bernadette Bensaude-Vincent and Christine Blondel, 43–63. Ashgate, 2008.

Roberts, C. J. D., ed. "Astronomy by Charles Babbage: A Series of Lectures Delivered at the Royal Institution in the Spring of 1815, Compiled from a Manuscript in the British Library." Unpublished printout, Royal Institution of Great Britain, 1989.

Ross, Sydney. "Scientist: The Story of a Word." *Annals of Science* 18, no. 2 (1962): 65–82.

"The Rotation of the Earth." *Illustrated London News* 18, no. 481 (May 3, 1851): 345 (ILNHA GPS).

"The Rotation of the Earth." *Illustrated London News* 18, no. 485 (May 17, 1851): 419–20 (ILNHA GPS).

Rothman, Patricia. "Ferguson, James (1710–1776)." *Oxford Dictionary of National Biography.* https://doi.org/10.1093/ref:odnb/9320.

"The Royal Colosseum." *Morning Chronicle*, December 25, 1856 (BLN GPS).

"The Royal Colosseum." *Morning Chronicle*, December 6, 1857 (BLN GPS).

Royal Museums Greenwich. "Lantern Slide." Inventory no. AST1098. https://www.rmg.co.uk/collections/objects/rmgc-object-11260.

Royal Museums Greenwich. "Orrery." Inventory no. AST1055. https://www.rmg.co.uk/collections/objects/rmgc-object-11217.

Royal Museums Greenwich. "Orrery." Inventory no. AST1062. https://www.rmg.co.uk/collections/objects/rmgc-object-11224.

Royal Museums Greenwich. "Orrery." Inventory no. AST1066. https://www.rmg.co.uk/collections/objects/rmgc-object-11228.

"Royal Polytechnic." *Theatrical Journal* 32, no. 1631 (March 15, 1871): [81] (BP PQ).

"Royal Polytechnic Institution: Lectures on Astronomy." *Critic*, March 6, 1847.

Royle, Edward. "Mechanics' Institutes and the Working Classes, 1840–1860." *Historical Journal* 14, no. 2 (June 1971): 305–21.

Rudwick, Martin J. S. *Earth's Deep History: How It Was Discovered and Why It Matters.* University of Chicago Press, 2014.

Rudwick, Martin J. S. *The Great Devonian Controversy: The Shaping of Scientific Knowledge Among Gentlemanly Specialists.* University of Chicago Press, 1985.

Rudwick, Martin J. S. *Scenes from Deep Time: Early Pictorial Representations of the Prehistoric World.* University of Chicago Press, 1992.

Ruiz-Castell, Pedro. "Astronomy and Its Audiences: Robert Ball and Popular Astronomy in Victorian Britain." *Antiquarian Astronomer*, no. 1 (January 2004): 34–39.

Ruse, Michael. "Evolution and the Idea of Social Progress." In Alexander and Numbers, *Biology and Ideology*, 247–75.

Ruse, Michael. "Introduction." In *Of the Plurality of Worlds: A Facsimile of the First Edition of 1853; Plus Previously Unpublished Material Excised by the Author Just Before the Book Went to Press; and Whewell's Dialogue Rebutting His Critics, Reprinted from the Second Edition, edited by Michael Ruse. University of Chicago Press, 2001.

Russell, Colin A. *Science and Social Change: 1700–1900*. Macmillan, 1983.

Sagan, Carl. *Cosmos*. Random House, 1980.

Sagan, Carl. "The Planets: The Earth as a Planet (1977)." Royal Institution. https://www.rigb.org/explore-science/explore/video/planets-earth-planet-1977.

Schaffer, Simon. "Natural Philosophy and Public Spectacle in the Eighteenth Century." *History of Science* 21, no. 1 (March 1983): 1–43.

Schaffer, Simon. "The Nebular Hypothesis and the Science of Progress." In *History, Humanity and Evolution: Essays for John C. Greene*, edited by James R. Moore, 131–64. Cambridge University Press, 1989.

"Science Gossip." *Athenaeum*, no. 2300 (November 25, 1871): 692 (BP PQ).

Science Museum Group. "Handbill: Bird." Inventory no. 1980-930/4. https://collection.sciencemuseumgroup.org.uk/objects/co8012680/handbill-bird.

Science Museum Group. "Mechanical Lantern Slide of the Solar System." Inventory no. 1902-104. https://collection.sciencemuseumgroup.org.uk/objects/co56380/mechanical-lantern-slide-of-the-solar-system.

Science Museum Group. "Orrery Made by John Rowley for the Earl of Orrery." Inventory no. 1952-73. https://collection.sciencemuseumgroup.org.uk/objects/co56970/orrery-made-by-john-rowley-for-the-earl-of-orrery.

Science Museum Group. "Orrery Planetary Model." Inventory no. 1869-48. https://collection.sciencemuseumgroup.org.uk/objects/co56210/orrery-planetary-model.

Science Museum Group. "Orrery Planetary Model by Benjamin Martin, 1738–1777." Inventory no. 1919-463. https://collection.sciencemuseumgroup.org.uk/objects/co56626/orrery-planetary-model-by-benjamin-martin-1738-1777.

Science Museum Group. "Orrery Planetary Model Designed by William Pearson, 1813–1822." Inventory no. 1950-55. https://collection.sciencemuseumgroup.org.uk/objects/co56968/orrery-planetary-model-designed-by-william-pearson-1813-1822.

Secord, Anne. "Botany on a Plate: Pleasure and the Power of Pictures in Promoting Early Nineteenth-Century Scientific Knowledge." *Isis* 93, no. 1 (March 2002): 28–57.

Secord, James. "How Scientific Conversation Became Shop Talk." In Fyfe and Lightman, *Science in the Marketplace*, 23–59.

Secord, James. "Knowledge in Transit." *Isis* 95, no. 4 (December 2004): 654–72.

Secord, James. "Monsters at the Crystal Palace." In *Models: The Third Dimension of Science*, edited by Soraya de Chadarevian and Nick Hopwood, 138–69. Stanford University Press, 2004.

Secord, James. "Newton in the Nursery: Tom Telescope and the Philosophy of Tops and Balls, 1761–1838." *History of Science* 23, no. 2 (June 1985): 127–51.

Secord, James. "Quick and Magical Shaper of Science." *Science* 297, no. 5587 (September 6, 2002): 1648–49.

Secord, James. *Victorian Sensation: The Extraordinary Publication, Reception, and Secret Authorship of "Vestiges of the Natural History of Creation."* University of Chicago Press, 2000.

Secord, James. *Visions of Science: Books and Readers at the Dawn of the Victorian Age*. Oxford University Press, 2014.

Shapin, Steven. "The Invisible Technician." *American Scientist* 77, no. 6 (1989): 554–63.

Shapin, Steven, and Simon Schaffer. *Leviathan and the Air-Pump: Hobbes, Boyle, and the Experimental Life*. Princeton University Press, 1985.

Shaw, Philip. *The Sublime*, 2nd ed. Routledge, 2017.

Sheets-Pyenson, Susan. "Popular Science Periodicals in Paris and London: The Emergence of a Low Scientific Culture, 1820–1875." *Annals of Science* 42, no. 6 (1985): 549–72.

Shepard, Elizabeth. "The Magic Lantern Slide in Entertainment and Education, 1860–1920." *History of Photography* 11, no. 2 (1987): 91–108.

Shteir, Ann B. *Cultivating Women, Cultivating Science: Flora's Daughters and Botany in England, 1760 to 1860*. Johns Hopkins University Press, 1996.

Smiles, Samuel. *Self-Help: With Illustrations of Character and Conduct*. London: John Murray, 1859.

Smith, Asa. *Smith's Illustrated Astronomy, Designed for the Use of the Public or Common Schools in the United States*. New York: Daniel Burgess & Co., 1855.

Smith, Crosbie, and Jon Agar, eds. *Making Space for Science: Territorial Themes in the Shaping of Knowledge*. Macmillan, 1998.

Smith, John Pye. *On the Relation Between the Holy Scriptures and Some Parts of Geological Science*. London: Jackson & Walford, 1839.

Smyth, C. P. "Account of the Astronomical Experiment on the Peak of Teneriffe in 1856." *Notices of the Proceedings at the Meetings of Members of the Royal Institution of Great Britain* 2 (March 5, 1858 [1854–58]): 493–97.

"The Solar Eclipse of July 8, 1842." *London Saturday Journal* 4, no. 79 (July 2, 1842): 23–24.

Sorrenson, Richard. "George Graham, Visible Technician." *British Journal for the History of Science* 32, no. 2 (June 1999): 203–21.

Steele, Richard. *The Englishman: Being the Sequel of the Guardian*. London: Printed by Sam Buckley in Amen-Corner, 1714.

Stephens, Michael D., and Gordon W. Roderick. "Science, the Working Classes and Mechanics' Institutes." *Annals of Science* 29, no. 4 (1972): 349–60.

Stephens, W. B. *Education in Britain: 1750–1914*. Macmillan, 1998.

Stevenson, Brian. "W. E. and F. Newton; Newton & Co." www.microscopist.net/NewtonCo.html.

Stewart, Larry. *The Rise of Public Science: Rhetoric, Technology, and Natural Philosophy in Newtonian Britain, 1660–1750*. Cambridge University Press, 1992.

A Stranger in London. "A Visit to St. Paul's." *Examiner*, March 7, 1824 (BP PQ).

Strutt, J. G., ed. *Tallis's History and Description of the Crystal Palace, and the Exhibition of the Worlds Industry in 1851*, vol. 2. London: John Tallis & Company, 1852.

"Sudbury." *Essex Standard*, April 30, 1841 (BLN GPS).

"Swimming: Deaths by Drowning." *Mirror of Literature, Amusement and Instruction* 6, no. 156 (August 20, 1825): 132–33.

Talfourd, Thomas N., ed. *The Letters of Charles Lamb, with A Sketch of His Life*. London: Edward Moxon, 1849.

Taub, Liba. "The 'Grand' Orrery." Inventory no. 1275. Whipple Museum of the History of Science, University of Cambridge, 2006. https://www.whipplemuseum.cam.ac.uk/explore-whipple-collections/astronomy/grand-orrery.

Taylor, Katie. "Mogg's Celestial Sphere (1813): The Construction of Polite Astronomy." *Studies in History and Philosophy of Science* 40, no. 4 (December 2009): 360–71.

Taylor, William C. "Polytechnic Schools in Manufacturing Districts." *North of England Magazine and Bradshaw's Journal of Politics, Literature, Science and Art* 1, no. 2 (March 1842): 75–83 (BP PQ).

"Theatres, &c.: The Past Week and the Coming One Dramatically Considered." *Era*, March 23, 1856.

Thomas, Keith. *In Pursuit of Civility: Manners and Civilization in Early Modern England*. Yale University Press, 2020.

BIBLIOGRAPHY

Thwaite, Ann. *Glimpses of the Wonderful: The Life of Philip Henry Gosse*. Faber and Faber, 2002.

Timbs, John. *Curiosities of London: Exhibiting the Most Rare and Remarkable Objects of Interest in the Metropolis*. London: J. C. Hotten, 1867.

Topham, Jonathan. "Beyond the 'Common Context': The Production and Reading of the Bridgewater Treatises." *Isis* 89, no. 2 (June 1998): 233–62.

Topham, Jonathan. "Biology in the Service of Natural Theology: Paley, Darwin, and the Bridgewater Treatises." In Alexander and Numbers, *Biology and Ideology*, 88–113.

Topham, Jonathan. "Publishing 'Popular Science' in Early Nineteenth-Century Britain." In Fyfe and Lightman, *Science in the Marketplace*, 135–68.

Topham, Jonathan. *Reading the Book of Nature: How Eight Best Sellers Reconnected Christianity and the Sciences on the Eve of the Victorian Age*. University of Chicago Press, 2022.

Topham, Jonathan. "Science and Popular Education in the 1830s: The Role of the 'Bridgewater Treatises.'" *British Journal for the History of Science* 25, no. 4 (December 1992): 397–430.

"Town Talk and Table Talk." *Illustrated London News* 19, no. 525 (October 18, 1851): 491 (ILNHA GPS).

[Untitled.] *Standard*, October 3, 1851 (BLN GPS).

Vanhoutte, Kurt. "Performing Astronomy: The Orrery as Model, Theatre, and Experience." In *Media Archeology and Intermedial Performance: Deep Time of the Theatre*, edited by Nele Wynants, 145–72. Palgrave Macmillan Cham, 2019.

Vanhoutte, Kurt, and Charlotte Bigg. "On the Border Between Performance, Science, and the Digital: The Embodied Orrery." *International Journal of Performance Arts and Digital Media* 10, no. 2 (September 2014): 255–60.

"Varieties: Mr. Adams' Lectures." *Literary Gazette*, no. 791 (March 17, 1832): 172 (BP PQ).

"Varieties: Mr. Adams's Orrery." *Literary Gazette*, no. 1314 (March 26, 1842): 221 (BP PQ).

"View of Roslyn Chapel, at the Diorama." *Mirror of Literature, Amusement, and Instruction* 7 (1826): 129–32.

Vincent, David. *Literacy and Popular Culture: England 1750–1914*. Cambridge University Press, 1989.

"A Voice from Egyptian Hall." *Punch* 11 (1846): 64.

Wach, Howard M. "Culture and the Middle Classes: Popular Knowledge in Industrial Manchester." *Journal of British Studies* 27, no. 4 (October 1988): 375–404.

BIBLIOGRAPHY

Walker, Adam. *Syllabus of a Course of Lectures, on Natural and Experimental Philosophy.* Liverpool: Printed by W. Nevett, and Co. in Princes-Street, 1771.

Walker, Deane F. *An Epitome of Astronomy, with the New Discovered Planets, and the Late Comet, as Illustrated by the Eidouranion,* 31st ed. Edinburgh: Printed by T. Colquhoun for the author, 1824.

Walker, William. *An Account of the Eidouranion, or the Transparent Orrery,* 10th ed. Bury St. Edmunds: Printed by P. Gedge, 1793.

Wallis, John. *A Brief Examination of the Nebulous Hypothesis, with Strictures on a Work Entitled Vestiges of the Natural History of Creation.* London: R. Groombridge & Sons, 1845.

Walters, Alice N. "Conversation Pieces: Science and Politeness in Eighteenth-Century England." *History of Science* 35, no. 2 (June 1997): 121–54.

Walters, Alice N. "Tools of Enlightenment: The Material Culture of Science in Eighteenth-Century England." PhD diss., University of California, Berkeley, 1992.

Warren, Samuel. *The Lily and the Bee: An Apologue of the Crystal Palace.* Edinburgh: William Blackwood and Sons, 1851.

"W. D. Saull." *Poor Man's Guardian,* November 24, 1832 (BLN GPS).

Weeden, Brenda. *The Education of the Eye: History of the Royal Polytechnic Institution 1838–1881.* Granta Editions, 2008.

Wellbeloved, Horace. *London Lions for Country Cousins and Friends About Town Being All the New Buildings, Improvements and Amusements in the British Metropolis.* London: William Charlton Wright, 1826.

Wells, Kentwood D. "Fleas the Size of Elephants: The Wonders of the Oxyhydrogen Microscope." *Magic Lantern Gazette* 29, nos. 2/3 (2017): 7–11.

Whewell, William. *Astronomy and General Physics Considered with Reference to Natural Theology,* 7th ed. London: William Pickering, 1839.

Wigelsworth, Jeffrey R., *Selling Science in the Age of Newton: Advertising and the Commoditization of Knowledge.* Ashgate, 2010.

Wilson, George, and Archibald Geikie. *Memoir of Edward Forbes.* London: MacMillan & Edmonston Co., 1861.

Wintroub, Michael. "Taking a Bow in the Theater of Things." *Isis* 101, no. 4 (December 2010): 779–93.

Withers, Charles W. J., and David N. Livingstone. "Thinking Geographically About Nineteenth-Century Science." In Livingstone and Withers, *Geographies of Nineteenth-Century Science,* 1–19.

Wood, H. T. "Bachhoffner, George Henry (1810–1879)," revised by M.

BIBLIOGRAPHY

C. Curthoys. *Oxford Dictionary of National Biography.* https://doi.org/10.1093/ref:odnb/982.

Wood, R. Derek. "The Diorama in Great Britain in the 1820s." *History of Photography* 17, no. 3 (October 1993): 284–95.

Wyld, James. *Notes to Accompany Mr. Wyld's Model of the Earth, Leicester Square.* London: Model of the Earth, 1851.

Yale Center for British Art. "Elton's Miniature Transparent Orrery, 1817." Inventory no. GV1199 E47. https://collections.britishart.yale.edu/catalog/orbis:580661.

Yanni, Carla. *Nature's Museums: Victorian Science and the Architecture of Display.* Johns Hopkins University Press, 1999.

Yeo, Richard. "Genius, Method, and Morality: Images of Newton in Britain, 1760–1860." *Science in Context* 2, no. 2 (Autumn 1988): 257–84.

Index

Note: Page numbers in **bold** indicate tables, page numbers in *italics* indicate figures, and references following "n" refer notes.

Abbott, Benjamin, 50
"Account of the Astronomical Experiment on the Peak of Teneriffe in 1856" (Smyth), 121–22
Account of the Eidouranion, or the Transparent Orrery, An (Walkers), 33–35
Adams, Charles Henry, 16, 46, 82–83, 109, 179, 230, 237; admission fees, 198–200, **199**, 214; advertisements for lectures by, 85–86, 93, *125*, 126–27, 182–83; audience of, 87–88; celebrity status of, 84–85; continuity of performances of, 85; early life and educational background of, 85–86; end of career of, 88–89; legacy of, 89–90, 226; and Newtonian science, 119; offering discounts, 200; reception of lectures of, 60, 87; reflecting popular taste, 128–29; rivalry with Bachhoffner, 90–96; and separation between professional and amateur practitioners, 96–103; showing solar system, 144; and solar eclipses, 119–20; stepping into lecturing business, 86–87; subjects of, 114–18, **116**, **117**, 150, 186; stage sideline of, 54–55; theaters used by, 51, 53, 86–89, 114–15; *Theatrical Observer* recommending, 60; and transparent orrery, 171, 173; vandalism of apparatus of, 86–87, 176; Vulcan-Mercury lecture of, 123–27, *125*. *See also* Bachhoffner, George Henry
Adams, George, 107, 154, 167
Address to Men of Science, An (Carlile), 220
Adelaide Gallery, 13, 37, 46, 89, 91, 96. *See also* Royal Polytechnic Institution
Adelphi Theatre, 54, 86, 88, 93, 198
affiliations (of lecturers), 5–6, 16, 168, 230–31; C. H. Adams as private showman, 83–90; motivations for entering astronomy lecturing business, 103–9; overview, 81–83, 109–10; scientific institutions, 90–96; separation between professional and amateur practitioners, 96–103. *See also* private entrepreneurs; private

INDEX

lecturing; public lecturing; scientific institutions
Airy, George Biddell, 4–5, 16, 216, 231, 238; and Friday Evening Discourses, 66, 67–68; presenting new discoveries, 120–22; tenure of, 98–100; transit of Venus, 127–28; visiting Ipswich, 78
Altick, Richard, 13, 37, 181, 213
amateur, term, 81–83, 102–3, 241n5
American Philosophical Society, 31
Analysis of Mr. Arden's Course of Lectures on Natural and Experimental Philosophy (Arden), 24
"Ancient and Scriptural Astronomy" (Beechey), 140–41
Anglican Church, 37. *See also* Church of England
apparatus, use of, 4, 7, 15–17, 87, 93, 108; astronomical lantern slides, 119, 181–84, 184; Diorama, 180–81; geography of lectures, 51, 53, 62, 72; Great Exhibition of 1851, 162–71; lecture subjects, 118, 123, 133, 150; magic lantern, 180–83; mechanical lantern slides, 183–85; orrery introduction, 153–62; overview, 151–53; pioneers of astronomy lecturing, 19–22, 24, 27, 32; prints of astronomical diagrams and illustrations, 185–87, 187; transparent orrery, 171–80. *See also* astronomical lantern slides; eidouranion; orrery; transparent orrery
Arago, François, 120
architecture, astronomy lecturing and, 37–38
Arden, James, 24
Arden, John, 24, 27
Armstrong, William, 93
Arnold, Samuel James, 57, 59, 173; and *Ouranologia*, 129–34, 173; and plurality of worlds, 148
artisan lecturers, 76, 83, 105

Ashted Working Men's Association, 73–74
Astronomer Royal (title). *See* Airy, George Biddell
"Astronomers and Astrologers" (Carlile), 220
astronomical lantern slides, 119, 181–84, 184
astronomical lectures. *See* astronomy lecturing
Astronomion, 173–74. *See also* eidouranion; Henderson, Ebenezer
"Astronomy: The Scale of the Solar System" (Strutt), 105
Astronomy Explained upon Sir Isaac Newton's Principles (Ferguson), 8, 31, 131–32, 134
astronomy lecturing: astronomy as "sublime science," 7–10; exploring wide spectrum of, 6–7; keen observation on, 3–4; legacies, 225–29; mapping uncharted terrain, 229–32; nineteenth-century science as contested space, 14–17; popular science and, 5–6; science in marketplace, 10–14; search for the sublime, 232–35; theatrical turn of, 6–7. *See also various entries*
Athenaeum (journal), 112, 126, 225
attendance, motivations for, 68, 78, 121, 197; astronomy fever, 214–16; "Deformito-mania," 213–14; implicit religious messaging, 209–10; intellectual and artistic fashion, 210–12; rational amusement, 213; root of polite culture of conversation, 213; Superficial Bore, 212–14. *See also* audiences (for astronomy lecturing)
"At the Play" (Hollingshead), 204–8, 216–17
audiences (for astronomy lecturing), 5–7, 224, 227; apparatuses used for attracting, 153, 177, 179, 183, 187; appealing to broader audiences,

INDEX

12–13; attendance records, 248n44; attracting, 23–24; balance between space and size, 53; capturing attention of, 26–27, 27; cheap lectures for, 73; composition of, 193–204; gallery audiences, 48–49; interpreting role of, 15; and lecture location, 37–39, 44, 48–51, 53–54, 60, 63–64, 68–73, 77–78, 80; motivations for attendance, 209–16; overview, 191–93; polarization of truth, 216–24; private entrepreneurs influencing, 231–33; theatergoer accounts, 204–9; and transparent orrery, 33; upper-class audience, 70, 193; Walkers pleasing, 10; working-class audiences, 71–73, 88, 222. *See also* attendance, motivations for; composition of audiences

Automatical Theatre, 50

BAAS. *See* British Association for the Advancement of Science
Babbage, Charles, 11, 66–67, 104–5, 238
Bachhoffner, George Henry, 16, 82–83, 109, 123, 231, 238; advertisements for lectures by, 92–95, 182–83; biography of, 90–91; and Colosseum (Regent's Park), 94–96; reviewing lectures of, 94; rivalry with C. H. Adams, 93–94, 200; Royal Polytechnic Institution and, 91–96; and separation between professional and amateur practitioners, 96–103; treatises published by, 90–91. *See also* Adams, Charles Henry
Baden-Powell, Robert, 65
Baily, Francis, 98, 113, 120
Ball, Robert Stawell, 4, 189, *190*, 228–29
Barkas, T. P., 78, 238
Bartley, George, 16, 46, 51, 86, 120, 173, 238; apparatuses used by, 171, 173, 176–77; commercial and aesthetic achievements of, 179; content of lectures of, 116–18, **116**, **117**, 148–49; delivering lectures on astronomy, 55, 59, 101; and *Ouranologia*, 129–34, 138; playbill of lecture of, 58; and playwrights, 129–31; and plurality of worlds, 148; portrait of, 56; and religious reflection, 140; reputation of, 53, 101, 229; theaters used by, 51, 52, 53, 101; transparent scenes used by, 183, 186

Barton, Ruth, 230
Bath, lectures in, 23, 31, 62, 219
Beauty of the Heavens, The (Blunt), 116–18, **116**, **117**, 185–87, *187*, 203
Beazley, Samuel, 51
Beechey, St. Vincent, 140–41, 238
Bellon, Richard, 41
Berman, Morris, 70
Berry, William, 77, 238
biblical works, quoting, 138. *See also* religion, astronomy lecturing and
Bird, John, 83, 105–8, *106*, 229, 230, 238
Bird, Wendy, 174
Birmingham, 23, 33, 72, 74
Bishop, George, 98
Blunt, Charles F., **116**, **117**, 185–87, *187*, 203
Bonnycastle, John, 25, 132
books, identifying target audiences for, 193–95. *See also* audiences (for astronomy lecturing)
bores. *See* Superficial Bore
Boswell, James, 31
Boulton, Matthew, 32
boxes (theaters), 48–49
Boyle, Charles, 153
Boyle, Robert, 20
Brande, William T., 68–69
Brewster, David, 131, 147, 165, 179–80, 189
Bridgewater Treatises, 12, 103, 221, 230; employing language and themes, 87; marble pellet analogy in, 139; and nebular hypothesis, 142–43; and

307

INDEX

plurality of worlds, 147; readership and, 192–93, 210; volume by Buckland, 210; volume by Whewell, 139, 142–43, 146–47, 230

Bristol, 23, 31

British Association for the Advancement of Science (BAAS), 11, 66, 72, 145, 203–4

Brooke, John, 138–39

Brougham, Henry, 10–11, 221

Buckland, William, 142, 210

Bullock, William, 45

Burke, Edmund, 9

Bury, J. B., 141

Busby, Charles A., 47

Butterworth, Mark, 174

Caledonian Theatre, 54

Callcott, John Wall, 117

Canterbury Cathedral, 44

Cantor, Geoffrey, 138–39

Cape of Good Hope, 99

career, term, 82–83

careers, pursuing, 103–9

Carlile, Richard, 218, 220–24

Cayley, George, 91

"Central Sun, The" (article), 3

Chalmers, Thomas, 146–47, 223, 238

Chambers, Robert, 11, 142

Chambers's Edinburgh Journal, 3, 11, 78

Chapman, Allan, 102, 134, 204, 214, 230; on adequately paid astronomical posts, 97–98; research by, 82

Chelmsford Mechanics' Institution, 72

Chemical Society of London, 99

Chemistry as Applied to the Fine Arts (Bachhoffner), 91

Children, Robert, 76, 83, 107–8, 231, 238

Christianity, scientific issues and, 15–16. *See also* religion, astronomy lecturing and

Christmas Lectures (Juvenile Lectures): Carl Sagan and, 151–52; John Wallis presenting, **65**, 68–69, 113–14; Michael Faraday presenting, 193–94, *194*, 196; as trademark attraction, 113. *See also* Royal Institution

Church of England, 37, 65

City of London, 54, 63–64. *See also* London, lecturing in

Clerke, Agnes Mary, 204, 227

clockwork, 47, 136, 154, 168, 173–74. *See also* apparatus, use of

Cole, Thomas, 131

Colladon, Jean-Daniel, 165

Collegiate Institution (Liverpool), 37

Colosseum (Regent's Park), 13, 44, 79, 90, 97, 102; and admission prices, 197–99, **199**; analyzing demise of, 37, 226, 231; astronomical lectures at, 84, 95, 123; and Bachhoffner, 94–96; financial crises of, 94–95, 226, 231. *See also* Bachhoffner, George Henry

Comic Annual, 214–16

commercial, term, 7, 14–16

commercial science, 7, 105; term, 6, 15, 103–4. *See also* sublime

composition of audiences: and admission fees, 198–200, **199**; audience categories, **196**; availability of sources, 197; difficulties of mapping, 193–94; finding clues about, 197–98; heterogeneous audiences, 197; for itinerant lectures, 197–98; juveniles and parents, 200–203, *202*; methods of surveying, 194–97; women, 203–4

Cornhill Magazine, 204–8

"Cosmical Philosophy" (Powell), 140

Cosmos (journal), 126

Cosmos (Sagan), 151, 234–35

Course of Experimental Philosophy, A (Desaguliers), 21

Covent Garden (theater), 42, 43, 46, 53, 59–60, 215; coverage of, 48; engagements at, 131; sideline at, 55

INDEX

Creation, renditions of, 54, 138–40
Creation, The (Haydn), 54
Critic, 94
Crop the Conjuror, 111
Crowe, Michael, 147
Cruickshank, George, 47, 48, 198
Crystal Palace (Hyde Park), 41, 163, 189; orreries at, 162–71, 180; post-Crystal Palace era, 227. See also Great Exhibition
Crystal Palace (Sydenham), 201, 202, 227
Curiosities of London (Timbs), 47, 49
Cyclopaedia (Rees), 179

Daguerre, Louis, 180
Daily Advertiser, 23
Dalton, John, 65, 238
Darwin, Charles, 61, 142
Davy, Humphry, 49, 63–64, 93, 203
"deficit model," adopting, 192
"Deformito-mania," 213–14
Demainbray, Stephen, 23
Derby, 23–24, 26, 27
Desaguliers, John Theophilus, 21–22, 28, 93, 154
Description and Use of an Orrery of a New Construction, The (Martin), 157
Description and Use of Both the Globes, Armillary Sphere and Orrery, The (Martin), 155
devotion. See religion, astronomy lecturing and
"Diagram of Ship." See Ouranologia (Arnold)
Dickens, Charles, 88, 179, 208
dioastrodoxon, 54, 131, 179
Dione (moon), 161
Diorama (at Regent's Park), 44, 180–81
"Discourse of Bores, A" (Poole), 212
discoveries, presenting news of. See scientific discoveries, presenting
Drury Lane Theatre, 42, 43, 48, 53, 60, 130–31

Earth (planet), 21, 64, 139, 148–49, 211; in Adams lectures, 87, 114–15; in Airy lectures, **66**, 128; in apparatus use, 151, 153–54, 158, 161, 170, 174–76, *184*, 186; and astronomical lantern slides, **116**, **117**; as conventional topic, 112; and debates on extraterrestrial life, 145; demonstrating tilt and rotation of, 30, 66–68, **66**, 67, 92, 105, 113; in Ouranologia, 132–34; and geological findings, 142; showing Earth-Moon system, 28, 30; in Walker lectures, 115; in Wallis lectures, 113–14
Earth's Crust, The (Page), 193
Edinburgh Encyclopaedia (Brewster), 179–80
Egyptian Hall (in Piccadilly), 41, 45–46, 213–14, 226. See also Bullock, William
eidouranion, 54, 55, 112, 204, 222, 225; first lecture on, 33–36; inspiring Samuel James Arnold, 130–31; inventing, 33, 171; merit of, 233; occupying English Opera House, 46–47; and plurality of worlds, 147; reproducing, 183; reviving, 182; scaled-down model, 177–79; sources on, 171–74, 176; success of, 36; technical methods of, 152. See also transparent orrery
eighteenth century: natural philosophy public lecturing in, 20–24; polite science culture in, 25–31; stage astronomy in, 31–36. See also nineteenth century, popular astronomy lectures in
"Electricity, Galvanism, and Other Branches of Natural Philosophy" (Bachhoffner), 92
Enceladus (moon), 161
English Opera House, 42, 43, 46, 53, 129, 230; admission prices for, 196, 198–200, **199**; delivering Lenten

INDEX

lectures on astronomy at, 55, 59; demonstrating eidouranion at, 54, 55, 130–31, 176, 204; facilities of, 51–52; playbill for lecture at, 58–59, *58*; praise for lectures at, 101. *See also* Lyceum Theatre

Epitome of Astronomy, An (Walker), 172–74

equipment. *See* apparatus, use of

Era (newspaper), 88, 93–94

Essex Standard, 108

Examiner, 13

extraterrestrials, considering existence of. *See* plurality of worlds

Eyton, John, 135–37

F. Plant of Nottingham, 167

Facey's Orrery, 165–67

facilities (of scientific institutions), lecturing in, 62–63, 70

Faraday, Michael, 5, 10, 49, 53, 70, 113, 212, 238; career start of, 13, 104, 109; Christmas Lecture of, 151–52, 193–94, *194*, 196; Friday Evening Discourse of, **66**, 68, 122–23; importance of showmanship, 50–51; on music, 53–54; and Royal Institution, 63–64, 90, 93; on Walker lectures, 53, 180

fashion of astronomy lecturing. *See* audiences (for astronomy lecturing)

Father and Son (Gosse), 208–9

Ferguson, James, 4, 8, 23, 27–28, 105, 107; accomplished popularizer of astronomy, 30–31, 36; apparatus of, 152, 158, 164, 170; influence of, 144, 147; *Ouranologia* and, 131–32, 134–35; playwright quoting, 148; Walkers and, 32. *See also Astronomy Explained upon Sir Isaac Newton's Principles* (Ferguson)

Fidler, John, 170

Finnegan, Diarmid, 39

Fontenelle, Bernard Le Bovier de, 146

Forbes, Edward, 61, 104

Francis, George William, 174–75

Freeman, Joseph, 77, 238

Friday Evening Discourse (series), 67–68, 120–23, 128; subjects covered in, **66**. *See also* Royal Institution

Frost, Isaac, 75–76, 222

Fulton, John, 168, 170–71

Fyfe, Aileen, 13, 195

Gallery of Nature, The (Milner), 8

gallery (theaters), 48–49

Genesis, geology challenging story of, 142–45

gentlemanly science, 79, 81, 97, 104, 110

Gentleman's Magazine, 68, 225

geography. *See* locations (of popular astronomy lecturing)

Geological Society, 61, 104

geology, 8, 14, 73, 111, 142, 193

George III (king), 23, 31, 154

Gieryn, Thomas, 39

Godwin, William, 119

Golinski, Jan, 9, 15, 152, 177–78

Gosse, Edmund, 208–9, 214

Graham, George, 21, 153–54

Grand Amateurs, 97–99, 102, 110, 231

grand orrery, painting of, 26–27, *27*

Grant, James, 48

Grant, Robert, 228, 238

Great Britian, lecturing in. *See various entries*

Great Exhibition, 13, 77, 180, 195, 227; showcasing visual aid at, 162–68. *See also* Crystal Palace (Hyde Park)

Great Globe (Wyld), 208–9

Great Metropolis, The (Grant), 48

"Greenwich-Cambridge Axis" network, 99, 231

Greenwich Royal Observatory, 98–99

Griffis, William, 22, 23, 32

Gross, Alan, 233–34

guidebooks, re-creating London with. *See London Lions for Country*

INDEX

Cousins and Friends About Town (Wellbeloved)

Hampshire Advertiser, 108
Handel, George Frideric, 54
Harris, John, 20
Hauksbee, Francis, 21–22, 93
Haymarket Theatre, 33, 42, 43, 48, 50, 54, 84, 88; admission prices for, 198–200, **199**
Hays, J. N., 83, 101, 109
Heavens Are Telling, The (Haydn), 54
Henderson, Ebenezer, 4, 6, 148–49, 173–74, 182, 239
Her Majesty's Theatre, 42, 43, 54, 88, 214
Herschel, Caroline, 203
Herschel, John, 4, 8, 98, 100, 165, 180; and nebular hypothesis, 144–45
Herschel, William, 31, 107, 142, 144
Higgitt, Rebekah, 203
"History of Modern Astronomy" (Beechey), 140
Hodgson, James, 20–22
Holden, Moses, 72, 239
Hollingshead, John, 204–9, 214
Holy, Holy, Lord (Handel), 54
Hood, Thomas, 59, 214–16
Hooke, Robert, 20
Howell, James, 54, 214, 239
Hutchins, Roger, 98–99
Huygens, Christiaan, 146, 153, 181
Hyde Park, 41, 162. *See also* Crystal Palace (Hyde Park); Great Exhibition
hydroelectric machine, 91–93

Iapetus (moon), 161
Illford Mutual Improvement Society, 77
Illustrated London News, 67, 68, 83, 92, 163
Inkster, Ian, 19, 83, 101, 229
institutional sites of analysis, 81, 253n2
Institution for the Diffusion of Knowledge, 72

instruments, public lectures and, 22. *See also* apparatus, use of
Introduction to Astronomy, An (Bonnycastle), 132, 161
invisible technician, 257n82
itinerant lecturers, 3, 23, 40, 70, 82, 86, 230; common practice of, 172; and eidouranion, 174, 177; flexibility of, 76–80; letter, 218

Johnson, Samuel, 31, 40
Jones, William, 158–61, *159*, *160*
"Jones-type" orrery, 158, *159*, 161, 167–68
Jupiter (planet), **117**, 148, 161
juveniles, astronomy lectures and, 93, **196**, *202*, 214, 224; in audience composition, 200–203; Ball lecture, 189–90, *190*; memories of visiting shows, 206–7; Wallis lectures, 113–14

Keevil, G. M., 174, *176*, 177, 183
Kemble, Stephen, 53
Kepler, Johannes, 119, 145, 164
Kew Gardens, 79
Kew Photoheliograph, 122
King, Henry C., 153, 167, 174
King's College London, 23, 104
King's Theatre, 42, 43, 46, 88; admission prices, 198–200, **199**
Kitchener, William, 59
Knight, Charles, 10–11
knowledge in transit, 38–39
Knowledge (journal), 228

Lach-Szyrma, Krystyn, 44, 48–49, 180, 247n36; memoir of, 40–42
Lacy, William, 161–62
Lamb, Charles, 201
Laplace, Pierre-Simon, 119, 137, 142, 145
Lardner, Dionysus, 4, 239
Lassell, William, 98
"Lecture in Lent" (cartoon), *206*, 207

311

INDEX

lecturers: artisan lecturers, 76, 83, 105; freelance lecturers, 68; Grand Amateurs, 97–99, 102, 110, 231; institutional lecturers, 6, 83, 101–2, 110, 112, 229; itinerant lecturers, 23, 40, 70, 76–80, 82, 86, 172, 230; motivations for entering astronomy lecturing business, 103–9; separation between professional and amateur practitioners, 96–103; private lecturers, 4–5, 68, 83, 85, 101, 103–4, 112; syllabi of, 113–20, **116, 117**; timeline related to, *100*, 101. *See also* itinerant lecturers; private entrepreneurs; private lecturing; public lecturing; scientific institutions; theaters, lecturing in

lectures. *See* astronomy, lectures on

Leeds, lecturing in, 70

Leeds Mercury, 89, 90

Leicester Square, 37, 41, 45, 209, 227

Leisure Hour (periodical), 107, 109

Lent, 4–5, 15, 38, 46, 120; amusements during, 83–84, 94, 129, *216*, 216–17; astronomy lecturing and, 53–55, 59–60, 69, 83–84, 92–95, 214, 232, 233; ban on regular plays during, 59–60, 130, 226–27; decline of astronomical lectures on, 225–29

Lescarbault, Edmond, 124, 126

Le Verrier, Urbain, 124–27, *125*

Lightman, Bernard, 5, 13, 14, 41, 79, 83, 229; illustrating prominent place of nonpractitioner popularizers, 230–31; warning of demarcation between popularizers and practitioners, 101–2

Lily and the Bee, The: An Apologue of the Crystal Palace of 1851 (Warren), 163–64

Lindley, John, 104

Linnean Society, 61

Linwood Gallery, 45, 89

lions, term. See *London Lions for Country Cousins and Friends About Town* (Wellbeloved)

Literary Gazette, 86–87, 112

Liverpool, lecturing in, 70

Liverpool Mechanics' Institution, 37, 70, 72

Liverpool Mercury, 70

Livingstone, David, 38

Lloyd, R. E., 54, 60, 138, 140, 179, **199**, 239

locations (of popular astronomy lecturing): itinerant lecturers, 76–80; lecturing beyond London, 69–76; London as lecturing location, 40–50; overview, 37–40; scientific institution consideration, 60–69; theater consideration, 50–60. *See also* London, lecturing in; scientific institutions; theaters, lecturing in

London, lecturing in, 61, **65, 66**, 94, *100*, 101; abundant resources offered by, 40–41; amusement attractions, 44–45; astronomy coverage, 46–47; guidebooks, 41–42, 44, 47, 96, 200; scientific institutions coverage, 49–50; theater coverage, 47–49. *See also* theaters, lecturing in

London Institution, 42, 43, 49, 61, 101; as imitator, 63–64; lecturers of, 68–69

London Lions for Country Cousins and Friends About Town (Wellbeloved), 41–42, 176; Adelaide Gallery in, 45–46; amusement attractions recommended by, 44–45; astronomy coverage, 46–47, 54, 200; humorous title of, 42; lantern technique according to, 181–82, 185; scientific institutions coverage, 49–50; theater coverage, 47–49

London Mechanics' Institution, 12, 42, 43, 49, 69, 71, **199**; audiences of, 197

London Saturday Journal, 75, 119

Love and Lunacy (Hood), 59, 214–16

312

INDEX

Lover of Truth, 218–23
Lunar Society, 26, 32
Lyceum Theatre, 51–54, *52*, 88, 130. See also English Opera House

Magazine of Science, 73, 174–75, *175*, 177
magic lantern, 72, 152, 174, 177, 189–90, *190*; and optical displays, 180–83
Magic Lantern, The: How to Buy and How to Use It (Brewster), 181
Manchester, lecturing in, 70, 71, 76, 140, 143
Manchester Literary and Philosophical Society, 65
marketplace, science in, 10–14
Martin, Benjamin, 4, 23, 33, 36, 171; apparatus used by, 153–59, *156*, 161–62, 167; as example of polite science culture, 27–31, *29*, *30*
Mathieu, Claude-Louis, 165
"Maximum Machine," 22
mean-motion orrery, 170. See also Pearson, William
Mécanique céleste (Laplace), 137
mechanical lantern slides, *176*, 177, 183–85, *184*
mechanics' institutes, 5, 76, 104, 197; audiences at, 70–73. See also Liverpool Mechanics' Institution; London Mechanics' Institution
Metropolitan Magazine, 87
Milky Way, **117**, 139
Millburn, John, 155, 158, 167
Milner, Thomas, 8
Mimas (moon), 161
Mirror, 74
Mirror of Literature, 44
Mogg's New Picture of London and Visitor's Guide to It [sic] Sight, 96
Moigno, Abbé, 126, 127
Monthly Notices of the Royal Astronomical Society, 124, 127, 228
Moon, 21, 87, 127, 132, 167, 170, 219; in *The Beauty of the Heavens*, 186–87, *187*; and educational lantern slides, 183; lectures on, 28–29, 210–11; 113–15, **116**, 120; orreries showing, 153, 158, 174
Moore, Daniel, 61–62
Moore, William P., 77, 239
Morning Chronicle, 86, 95
Morning Herald and Daily Advertiser, 34
Morrell, Jack, 97
Morton, Alan, 23
Morus, Iwan, 14
Mr. Facey. See Facey's Orrery
Muggleton, Lodowicke, 75
Muggletonians, 75–76, 222–23
music (in theaters), considering, 53–54. See also theaters, lecturing in
Music Room (Oxford), 54

Nall, Joshua, 6
National Maritime Museum, 158–61, *159*, *160*, 183–84, *184*
National Standard Theatre, 254n32
natural philosophy, public lecturing on, 62, 82, 150, 169, 185, 200, 228; branching out beyond London, 23; commercial nature of lectures, 22, 31; courses, 64, **65**; instrument orientation of curricula, 22, 24, 36; and Newtonian science, 132–33; and polite science culture, 26–28, 29; public lectures on, 12, 14, 19, 64–65; publishing accounts of lectures, 23–24; success creating followers, 23
"Natural Philosophy" (Hodgson), 20–21
natural theology, 9, 15, 200, 221, 224; Bridgewater Treatises and, 12, 103; Christianity and, 150, 222–23; idea and rhetoric of, 137–41; plurality and, 146–49
Natural Theology (Paley), 137–38
nebular hypothesis, 17, 103, 147, 150, 218, 221; and notions of progress, 141–45

INDEX

Neptune (planet), 114, **117**, 123–24, 140
New Monthly Magazine, 212
Newtonian science, 17, 19, 127, 141, 149, 233, 268n19; and astronomy, 25, 75–76, 222; international spread and consolidation of, 20–21; in lectures, **116**, 118–19; in *Ouranologia*, 131–33
Newton, Isaac, 8, 20–21, 27, 31, 93, 127; in lectures, 118–19, 132; prestigious status of, 118–19, 150; sublimity of, 164
Newton & Company, 167, 184, 265n42
Newton & Son, 167, 265n42
Nichol, John Pringle, 16, 72, 218, 221, 223, 228, 239; and nebular hypothesis, 143–45
Night-Thoughts (Young), 134–35, 137, 146
nineteenth century, popular astronomy lectures in: contested space, 14–17; legacies, 225–29; mapping uncharted terrain, 229–32; overview, 4–17; search for the sublime, 232–35. *See also various entries*
"nonpractitioner" popularizers, 5–6
Northern England, lecturing in, 69–70
Nurse, William Mountford, 91

Of the Plurality of Worlds: An Essay (Whewell), 147
"On Mr. De la Rue's Photographic Eclipse Results" (Faraday), **66**, 122–23
"On Rotatory Stability; and its Applications to Astronomical Observations on board Ships" (Powell), **66**, 128
Origin of Species (Darwin), 142
orrery, 3, 17, 22, 28, 34, 36; of Bartley, 58, *59*, 215–16; of C. H. Adams, 60, 84, 86, 89–90, 115, 123, *125*, 126, 161, 176; Charles Dickens describing, 208; conformity of manufacturing, 167–68; double-cone wheelwork mechanism, 155–57, *156*; evolution of, 187–90, *188*, *190*; Facey's Orrery, 165–67; grand orrery, 26, 27, 28, 153–54, 157–58; hybrid "new portable orrery," 158–61, *159*; hydraulic orrery, 47; iconic status of, 153; "improved Orrery for mean Motions," 170; "Jones-type" models, 158–61, *159*, 167–68, 182; lecturers using, 51, 95, 132, 154–55, 201, *202*; low evaluation of, 167, 179–80; mean-motion orrery, 170; more sophisticated models, 168–71; "Orrery for equated Motions in three Parts," 170; other orrery makers, 161–62, 165–67; portable orrery, 28–30, *30*, 158–59, *159*, 161, *166*, 167, 182; prototype design of, 153–54; recognizing educational value of, 21, 72, 213, 222; in reminiscences, 204–9; simplifying cost of, 157–58; term, 21; transparent orrery, 7, 10, 17, 33, 55, 77, 171–80, *175*, *178*, 183–85, 212, 214, 225; usage of, 151–52; vandalizing, 86–87; Vertical Orrery, 162–65. *See also* eidouranion; transparent orrery
Ouranologia (Arnold), 173, 233; association between astronomy and religion in, 134–38; inspiration for, 130–31; Newtonian science in, 132–33; overview of, 129–30; religious renditions of Creation in, 139–40; showing considerations of playwright, 133–34; subject arrangement in, 131–32
Oxford English Dictionary, 8, 82
Oxford University Herald, 108

Page, David, 193–94
Pale Blue Dot (Sagan), 151
Paley, William, 137–38
Pall Mall, 45
parents, astronomy lectures and, 200–203, *202*. *See also* audiences

INDEX

(for astronomy lecturing); composition of audiences
Parsons, William (Lord Rosse), 98, 223
Patterson, Robert, 131
Payne, Charles, 91
Pearson, William, 168–71, *169*, 179–80
"Pendulum-experiments lately made in the Harton Colliery, for ascertaining the mean Density of the Earth, The" (Airy), **66**, 128
Penny Cyclopaedia, 10–11, 211
Penny Magazine, 10–11, 73, 212, 215
people, bringing astronomy lecturing to: advertisement, 72–74; cheap books and periodicals, 73; engagements beyond London, 70–71; knowledge as readily available commodity, 73–74; mechanics' institutes, 70–72; quality of lectures, 74–76; working-class audiences, 72–73. *See also* audiences (for astronomy lecturing); composition of audiences
Pepper, John Henry, 95, 102, 239; restoring astronomical lectures, 226; reviewing astronomical lecture of, 89–90
phantasmagoria, 72, 181–83
Philosopher Giving That Lecture on the Orrery, A (Wright), 26–28
Philosophical Enquiry into the Origin of Our Ideas of the Sublime and Beautiful, A (Burke), 9
pioneers (of astronomy lecturing): overview, 19–20; natural philosophy and, 20–24; polite science, 25–31; stage astronomy, 31–36. *See also* natural philosophy; polite science culture; stage astronomy
pit (theaters), 47, 48–49
planetarium, 21–22, 174; at Great Exhibition, 165–70; of Mr. Perini, 204; and orreries, 153–55, 157–60, *160*; projectors, 234; of Roemer, 157; using, 28–30, *30*

Planetary Machine, 21
planetary motions, 69, 114–15, 119, 137, 168, 171
playwrights, astronomy lecturing and, 129–34
plurality of worlds, 141, 218, 222, 224; debating, 145–50
Podmore, J. B., 72–74, 239
polite science culture, 25–26, 200, 207, 211, 213, 224; Martin and Ferguson and, 28, 30–31; orrery as instrument of, 153–62; painting as demonstration in context of, 26–27; textbook as example of, 27–28, 29
Pond, John, 16, 65, **65**, 239
Poole, John, 212, 214–15
Popham, Charles, 60, 239
popular astronomy lectures. *See* astronomy lecturing
popular science, 9, 228–29, 233–34; astronomy as, 19, 25, 214; defense of, 6; demarcation between professional science and, 102; emergence of term, 12–14; emotional appeal of, 15; fashion and, 201, 203, 210–12, 214; historical studies of, 5–6; natural theology and, 134; 137; opportunities presented by, 103–4; publishing trade, 191–93, 195
Popular Treatise on Voltaic Electricity and Electro-Magnetism, A (Bachhoffner), 91
Porter, Roy, 82
Powell, Baden, 16, 67–68, *67*, 239; courses of, 64–65, **65**, 140–41; Friday Evening Discourses of, **66**, 67, 128; and natural theology, 138; and plurality of worlds, 147
Powers, Jonathan, 27
Presbyterian Church, 146–47
Priestley, Joseph, 32
Principia (Newton), 118
print culture, science and, 11–12
private entrepreneurs, 4–5, 7, 82, 101–2,

315

INDEX

109; drawing inspiration from other popular works, 129–34; presenting scientific discoveries, 123–27. *See also* Adams, Charles Henry

private lecturing, 4–5, 68, 83, 85, 129; dichotomy between institutional and private lecturers, 83, 101–3; everyday records of, 197–98; exploiting commercial science, 104; institutionalization eclipsing, 101; John Bird and, 107–8; overview of, 230–32; painting of, 26; publications of, 112. *See also* Adams, Charles Henry

Proctor, Richard Anthony, 127, 227–28
profession, term, 82–83
professional, term, 102
progress, science of, 141–45
prosperity, prospects for, 103–9
provincial towns, lecturers in, 76–80
public lecturing, 25, 99, 228; active involvement in, 65; foundations laid by, 36; growth of, 12; horizon of trade of, 31–32; monopoly on, 23; motivations for, 83; on natural philosophy, 20–24; orreries and, 21; as private business, 109; pursuing career in, 86. *See also* astronomy lecturing

Punch (magazine), 84–85, 123, 201, 210; cartoons published by, *202, 206, 207*; on public's taste for the "Monstrous," 213–14

quality (of lectures), 74–76, 207–8

Ranyard, Arthur Cowper, 227–28
Ratcliff, Jessica, 127–28
Real or Constitutional House That Jack Built (Hone), *136*, 137
recent occurrences, lectures and news of. *See* scientific discoveries, presenting
Rees's Cyclopaedia, 170, 179
Reeve, John, 75

Regency era, 5, 149, 198, 213, 230
Regent's Park, 13, 37, 44, 79, 94, 180, 197
Reid, Francis, 222
religion, astronomy lecturing and, 19, 37, 136; Carlile–Rogers dispute, 218–23; lecturing beyond Lent, 140–41; Lover of Truth, 218–23; natural theology, 137–39; promoting Christianity, 134–35; rendering Creation, 139–40
Religious Tract Society (RTS), 12, 103, 135, 191, 230
reminiscences. *See* theatergoers, accounts of
Republican (journal), 218–20, 223. *See also* Carlile, Richard; Lover of Truth
Rhea (moon), 161
rhetorical devices. *See* natural theology
Robinson, Thomas Romney, 223
Roemer, Ole, 153, 157
Rogers, A. F., 218–23, 240
Rosslyn Chapel, 44
Rowley, John, 153, 157
Royal Armoury and Automatons, 45
Royal Astronomical Society (RAS), 4, 61, 65, 98–99; contributions to, 169–70; graduates of, 227–28; *Monthly Notices of the Royal Astronomical Society*, 124, 127, 228
Royal Institution, 5, 12–13, 15, 61, 100, 150; admission prices, 198–200, **199**; annual subscription for lectures at, 70–71; apparatuses used at, 151, 169–70, 189; astronomy lecturing in theaters, 50–52; attendance at, 68–70; attendance records, 248n44; audiences at, 193–95, *194,* 197; and career prospects, 104–5; Carl Sagan lecturing at, 151–52; Christmas Lectures, 113, 115, 193, 196; courses at, 64–69, **65**, 228, 230–31, 234; dichotomy between institutional and private lecturers, 101–2; founding of, 10–11; Friday Evening Discours-

316

es, 66–68, **66**, 120–23; involvement in management of, 61–62; John Timbs writing about, 49; journals of, 112; juvenile lectures at, 113–14; in London, 38, 42, 43, 45; maintaining strong connections, 228; mechanics' institutes and, 70; Michael Faraday at, 50–51, 90, 93; motivation for visiting, 209–10; presenting new discoveries at, 120–21, 123, 128; "rational recreations" for science lovers, 41; religious reflection at, 140; as role model for providing institutional lectures, 63–64; Samuel James Arnold and, 130; snobbery of, 70–71; subscription fee of, 71; syllabi used at, 113, 116–18, **116, 117**, 140; William Pearson and, 169; women's participation at, 203; writing about, 49

Royal Manchester Institution, 71, 140, 143

Royal Polytechnic Institution, 13, 16, 101–2, 231; admission prices, 198–200, **199**; auditoriums of, 97; Bachhoffner at, 84, 90–96, *92*; fate of, 96; founding of, 46, 91; name change, 255n44; path between scientific use and profitable entertainment, 255n47; publicizing programs, 93–94; reviewing lectures at, 89–90; reviving astronomical lectures at, 226. *See also* Bachhoffner, George Henry

Royal Society, 11, 49, 64, 93, 109, 121; fellows of, 31, 65, 67, 169; and natural philosophy, 20–23

"Rudiments of Astronomy, The" (Wallis), 113

rural settlement, lecturers in, 76–80

Russell Institution, 63

Sagan, Carl, 113, 151–52, 189, 234–35
Salisbury, lecturing in, 23

Saull, William Devonshire, 73, 198, 239
Schaffer, Simon, 141
science in the marketplace, phrase. *See* marketplace, science in
Science Museum in London, 23
scientific discoveries, presenting, 63, 112; popular press, 127–29; private entrepreneurs, 123–27; scientific institutions, 120–23
scientific institutions, lecturing in, 94, 101, 109, 168, 197, 228, 230; admission prices, 198–200, **199**; affiliation with the Royal Polytechnic, 90–96, 101; astronomical lectures at the Royal Institution, 64–69, **65**, **66**; attendance records, 248n44; economic prosperity, 60–61; facility selection, 62–63; freelance lecturers, 68–69; Friday Evening Discourses, **66**, 67–68; locations, 44; *London Lions* coverage, 49–50; nebular hypothesis, 143–44; networking and research resources, 61–62; presenting latest discoveries, 127–28; regulations, 71; religious flavor, 140; Royal Institution as role model, 63–64; rural settlements, 78; spatial characteristics, 79–80; tracing roots of, 81–82
scientist, existence of profession of, 82–83
scientist, term, 11–12
SDUK. *See* Society for the Diffusion of Useful Knowledge
Secord, Anne, 213
Secord, James, 75, 195, 229; comparing architecture of institutions, 37–39; gentlemanly science, 97; knowledge in transit, 39; studying *Vestiges*, 192, 210–11; using term "commercial science," 6–7, 14, 103–4
self-improvement, potential prosperity and, 108–9, 166
Series of Discourses on the Christian

INDEX

Revelation, Viewed in Connection with the Modern Astronomy, A (Chalmers), 146
Seurat, Claude-Ambroise, 45
Shipwrecked Fishermen and Mariners' Benevolent Society, 77
Short Account of a Course of Natural and Experimental Philosophy, A (Arden), 24
shows and exhibitions, science and, 12–13
Shows of London, The (Altick), 13
sideline, stage astronomy as, 54–55, 59
size (of theaters), considering, 53
Smart, Benjamin, 53
Smatter, Sam, 212, 214–15
Smith's Illustrated Astronomy, 189
Smyth, Charles Piazzi, 67–68, 121–22, 239
Smyth, William Henry, 98, 128
"Society for Scientific, Useful, and Literary Information," 74
Society for the Diffusion of Useful Knowledge (SDUK), 10–11, 73, 103

solar eclipses, lecturers addressing, **66**, 68, 119–20, 128; presenting discoveries, 120–23
solar system, 111, 139, 176–77, 205, 209; apparatuses reflecting, 151, 153, 165, *176*, 180, 183, *188*, 189; lecturing on, 72, 105, 114–15, **116**, 132, 143, 151; mechanical models of, 21, 30; and nebular hypothesis, 142–44; "Order Is Heaven's First Law," 136–37, *136*; showing current state of, 144
Somerville, Mary, 203
South, James, 98
space, balance between audience size and, 53. *See also* theaters, lecturing in
spectators. *See* audiences (for astronomy lecturing)
stage astronomy, 14, 17, 46, 86, 180, 183, 229; emergence of, 31–36; faded

convention of, 226; and music, 53–54; in theaters, 80
Standard (newspaper), 88–89, 163–64, 166
Star-Land (Ball), 189, *190*, 228
Steele, Richard, 213
Stores of Knowledge, 211
Story of the Heavens, The (Ball), 228
Strand (street), 37, 48, 86
Strand (theater), 42, 43, 54
Strutt, John William, 105, 240
subjects (in astronomy lecturing): association between astronomy and religion, 134–41; drawing inspiration from other popular works, 129–34; latest scientific discoveries, 121–29; lecture syllabi, 113–20, **116**, **117**; overview, 111–12; plurality of worlds, 145–50; science of progress, 141–45
sublime, 34, 54, 59, 207, 214, 232; astronomy as "sublime science," 7–10, 36, 72, 80, 113, 150; evoking feelings of, 186–88; religion and, 103, 134–41; searching for, 232–35; term, 7–10, 15–16; in theaters, 50–51; tools of, 152; understanding, 137
Superficial Bore, 212–14
"Supernumerary Computers," 99
Surrey Institution, 42, 43, 63
syllabus, 34, 69, 73, 112, 148–49; Adams lectures, 114–15, **116**, **117**, 120; common subjects, 113–20, 186; finding clues from, 197; narratives of inhabited extraterrestrial worlds in, 147; nebular hypothesis in, 143–45; Newtonian science as key component in, 118–19; of Nichol, 143; *Ouranologia* as, 129–34; pioneers using, 21–23; of Powell, 140; solar eclipses in, 119–20; Walker subjects, 115–17, **116**, **117**; Wallis lectures, 113–14, 116–18, **116**, **117**

INDEX

Tallis's History and Description of the Crystal Palace, 164–65
Taylor, William Cooke, 72
technology and industry, advancement of, 11–12
Telegraph, 89
Tethys (moon), 161
theatergoers, accounts of, 204–9
theaters, lecturing in, 4, 7, 13, 79–80, 171; accounts of visiting shows, 204–9; advertisements, 112; audiences in Regency and Victorian eras, 198; auditorium differences, 97; ban on regular plays during Lent, 59–60, 130, 226–27; blending instruction with moral education and religious sentiments, 54; C. H. Adams using theaters, 88; considering theater size, 53; distinctive facilities, 51–52; divisions, 48–49; fashion, 209–10, 214; historians overlooking theaters, 39–40; juveniles, 200; lecture duration, 69; lecture organization, 51; Lent as season for astronomy lecturing, 60; locations of theaters, 42–44, 42, 43; *London Lions* coverage, 47–49; music considerations, 53–54; spatial features, 50; stage astronomy as sideline, 54–55, 59; West End theaters, 7, 38, 48, 97, 230, 234
Theatrical Journal, 89, 226
Theatrical Observer, 60, 86, 87
"theatrical turn," 7, 10, 14, 36, 80, 232, 234
Timbs, John, 47
Titan (moon), 161
Tompion, Thomas, 153
tools. *See* apparatus, use of
Topham, Jonathan, 12, 192–93
"Total Solar Eclipse of 1851, July 28, The" (Airy), 120–21
transparent orrery, 7, 10, 17, 33, 77; criticism of, 179–80; favorability of, 171–72; lack of surviving information of, 171; limits of, 183–85; "Miniature Transparent Orrery," 177–79, *178*, 183; popularity of, 31–36; revealing hybrid characteristics of, 174–77, *175*; and Superficial Bore, 212, 214; Walkers inventing, 33–34, 171–73, 225, 232
Treatise on Astronomy, A (Henderson), 148–49, 182
Treatise on Astronomy, A (Herschel), 180
truth, polarization of: Lover of Truth, 218–23; opinions of "At the Play," 216–17; reaction diversity, 217–18
truth-spots, 39
Two Systems of Astronomy (Frost), 76, 222–23
Tyndall, John, 5, 63, 113

United Kingdom. *See* Great Britain, astronomy lectures in
University of London, 49
University College London, 104
Uranographia (Blunt), 185
"Utilitarians," 70

venues. *See* locations (of popular astronomy lecturing); London, lecturing in; Royal Institution; scientific institutions; theaters, lecturing in
Vertical Orrery (Great Exhibition), 162–65
Vestiges of the Natural History of Creation (Chambers), 11, 142, 145, 192, 210–11
Victorian era, lecturing culture in. *See* astronomy lecturing
Victorian Popularizers of Science (Lightman), 229
Victorian Sensation (Secord), 229
Views of the Architecture of the Heavens (Nichol), 143, 145
visitors' reminiscences. See *London Lions for Country Cousins and Friends About Town* (Wellbeloved)

INDEX

visual aid. *See* apparatus, use of
Vulcan (hypothetical planet), 123–27

W. & S. Jones (company), 30, 119, 167, 182; and orreries, 158–61
Walker, Adam, 4, 10, 31–36, 35
Walker, Deane Franklin, 35, 46, 54, 55, 76, 182, 240; advertising objectives of, 138; death of, 225–26; inspiring Samuel James Arnold, 130–31; and Newtonian science, 119; publications of, 112; showing solar system, 144; subjects of, 114–18, **116**, **117**; taking over family lecturing business, 35–36
Walker, William, 10, 34, 35, 54, 80, 180, 240; introducing eidouranion, 33–35, 171–73; and plurality of worlds, 147–48; praise for, 51, 53; stage astronomy of, 31–36
Wallis, John, 16, 231, 240; as freelance lecturer, 68–70, 72, 82, 101, 109; lectures of, **65**, 113–14, 116–18, **116**, **117**; showing solar system, 144
Walters, Alice, 26
Warren, Samuel, 163–64
Wellbeloved, Horace, 42, 44, 46–47, 54
Wess, Jane, 23

Whewell, William, 8, 25, 118, 138, 139; coining the term "scientist," 11; describing nebular hypothesis, 142–43; describing nebular hypothesis, 230; and plurality of worlds, 146–47
Wigelsworth, Jeffrey, 22
William IV (king), 107
Withers, Charles, 38, 203
Wollstonecraft, Mary, 24
Wolverhampton, lecturing in, 23
women, astronomy lectures and, 203–4. *See also* audiences (for astronomy lecturing); composition of audiences
"Wonders of Astronomy, The" (Moore), 77
working-class audiences, astronomy lecturing to, 72–73
Wright, Joseph, 26–28, 27, 154
Wright, Thomas, 153

Young, Edward, 134, 137, 146
Young, Thomas, 64, **65**, 169, 240
Young Gentleman and Lady's Philosophy, The (Martin), 27–28, 29, 155, *156*
Youth Magazine, 223

Zeiss, Carl, 234